U.S. Marine Corps Close Combat

Plus

Military Working Dogs

Department of Defense

Pentagon
Publishing

DEPARTMENT OF THE NAVY
Headquarters United States Marine Corps
Washington, D.C. 20380-1775

18 February 1999

FOREWORD

1. PURPOSE

Today's Marines operate within a continuum of force where conflict may change from low intensity to high intensity over a matter of hours. Marines are also engaged in many military operations other than war, such as peacekeeping missions or noncombatant evacuation operations, where deadly force may not be authorized. During noncombative engagements, Marines must determine if a situation warrants applying deadly force. Sometimes Marines must decide in a matter of seconds because their lives or the lives of others depend on their actions. To make the right decision, Marines must understand both the lethal and nonlethal close combat techniques needed to handle the situation responsibly without escalating the violence unnecessarily. Marine Corps Reference Publication (MCRP) 3-02B, *Close Combat,* provides the tactics, techniques, and procedures of Marine Corps close combat. It also provides the doctrinal basis for the Marine Corps Close Combat Training Program (MCCCTP).

2. SCOPE

This publication guides individual Marines, unit leaders, and close combat instructors in the proper tactics, techniques, and procedures for close combat. MCRP 3-02B is not intended to replace supervision by appropriate unit leaders and close combat instruction by qualified instructors. Its role is to ensure standardization and execution of tactics, techniques, and procedures throughout the Marine Corps.

3. SUPERSESSION

MCRP 3-02B supersedes Fleet Marine Force Manual (FMFM) 0-7, *Close Combat,* dated 9 July 1993. There are significant differences between the two publications. MCRP 3-02B should be reviewed in its entirety.

4. WARNING

Techniques described in this publication can cause serious injury or death. Practical application in the training of these techniques will be conducted in strict accordance with approved Entry Level Close Combat, Close Combat Instructor (CCI), and Close Combat Instructor Trainer (CCIT) lesson plans. Where serious danger exists, the reac er is alerted by the following:

WARNING

5. CERTIFICATION

Reviewed and approved this date.

BY DIRECTION OF THE COMMANDANT OF THE MARINE CORPS

J. E. RHODES
Lieutenant General, U.S. Marine Corps
Commanding General
Marine Corps Combat Development Command

DISTRIBUTION: 14400006600

OVERVIEW OF CLOSE COMBAT

1. Purpose of Close Combat

Close combat is the physical confrontation between two or more opponents. It involves armed and unarmed and lethal and nonlethal fighting techniques that range from enforced compliance to deadly force. The purpose of close combat is to execute armed and unarmed techniques to produce both lethal and nonlethal results. Unarmed techniques include hand-to-hand combat and defense against hand-held weapons. Armed techniques include techniques applied with a rifle, bayonet, knife, baton, or any weapon of opportunity.

2. Continuum of Force

Marines will find themselves in both combative and noncombative situations. The threat level in these situations can rise and fall several times based on the actions of both Marines and the people involved. The escalation of force stops when one person complies with the demands imposed by another person. This range of actions is known as a continuum of force. Continuum of force is the concept that there is a wide range of possible actions, ranging from voice commands to application of deadly force, that may be used to gain and maintain control of a potentially dangerous situation (MCO 5500.6_, *Arming of Security and Law Enforcement [EE] Personnel and the Use of Force)*. Continuum of force consists of five levels that correspond to the behavior of the people involved and the actions Marines use to handle the situation (see the table below). Close combat techniques are executed in levels three, four, and five.

Level One: Compliant(Cooperative)

The subject complies with verbal commands. Close combat techniques do not apply.

Level Two: Resistant (Passive)

The subject resists verbal commands but complies immediately to any contact controls. Close combat techniques do not apply.

Continuum of Force		
Level	**Description**	**Actions**
1	Compliant (Cooperative)	Verbal commands
2	Resistant (Passive)	Contact controls
3	Resistant (Active)	Compliance techniques
4	Assaultive (Bodily Harm)	Defensive tactics
5	Assaultive (Serious Bodily Harm/ Death)	Deadly force
Note: Shading indicates levels in which Marines use close combat techniques.		

Level Three: Resistant (Active)

The subject initially demonstrates physical resistance. Marines use compliance techniques to control the situation. Level three incorporates close combat techniques to physically force a subject to comply. Techniques include—

- Come-along holds.
- Soft-handed stunning blows.
- Pain compliance through joint manipulation and the use of pressure points.

Level Four: Assaultive (Bodily Harm)

The subject may physically attack Marines, but he does not use a weapon. Marines use defensive tactics to neutralize the threat. Defensive tactics include the following close combat techniques:

- Blocks.
- Strikes.
- Kicks.
- Enhanced pain compliance procedures.
- Nightstick blocks and blows.

Level Five: Assaultive (Serious Bodily Harm/Death)

The subject usually has a weapon and will either kill or seriously injure someone if he is not stopped immediately and brought under control. Typically, to control the subject, Marines apply deadly force through the use of a firearm, but they may also use armed and unarmed close combat techniques.

3. Marine Corps Tactical Concepts

Close combat techniques support the following key Marine Corps tactical concepts. The concepts are not standalone ideas but are to be combined to achieve an effect that is greater than their separate sum.

Achieving a Decision

Achieving a decision is important in close combat. An indecisive fight wastes energy and possibly Marines' lives. Whether the intent is to control an opponent through restraint or defend themselves in war, Marines must have a clear purpose before engaging in close combat and act decisively once engaged.

Gaining an Advantage

A basic principle of martial arts is to use the opponent' s strength and momentum against him to gain more leverage than one's own muscles alone can generate, thereby gaining an advantage. In close combat, Marines must exploit every advantage over an opponent to ensure a successful outcome. This can include employing various weapons and close combat techniques that will present a dilemma to an opponent. Achieving surprise can also greatly increase leverage. Marines try to achieve surprise through deception, stealth, and ambiguity.

Speed

Marines use speed to gain the initiative and advantage over the enemy. In close combat, the speed and violence of the attack against an opponent provides Marines with a distinct advantage. Marines must know and understand the basics of close combat so they can act instinctively with speed to execute close combat techniques.

Adapting

Close combat can be characterized by friction, uncertainty, disorder, and rapid change. Each situation is a unique combination of shifting factors that cannot be controlled with precision or certainty. For example, a crowd control mission may call for Marines to employ various techniques ranging from nonlethal restraint to more forceful applications. Marines who adapt quickly will have a significant advantage.

Exploiting Success

Typically, an enemy will not normally surrender simply because he was placed at a disadvantage. Marines cannot be satisfied with gaining an advantage in a close combat situation. They must exploit any advantage aggressively and ruthlessly until an opportunity arises to completely dominate the opponent. Marines must exploit success by using every advantage that can be gained.

(reverse blank)

Close Combat

Table of Contents

CHAPTER 1

FUNDAMENTALS OF CLOSE COMBAT

This chapter describes all techniques for a right-handed person. However, all techniques can be executed from eittwr side.

The Marine is depicted in camouflage utilities. The opponent is depicted without camouflage.

The fundamentals of close combat include ranges, weapons of the body, target areas of the body, and pressure points of the body. These fundamentals form the basis for all close combat techniques. They provide Marines with a common framework regardless of the type of confrontation or the techniques used. If Marines apply these fundamentals properly in a close combat situation, they may save their lives or the lives of fellow Marines.

1. Ranges of Close Combat

Close combat engagements occur within three ranges: long range, midrange, and close range. During any engagement, these ranges may blur together or may rapidly transition from one to another until either the opponent is defeated or the conflict is resolved.

Long Range

During long range engagements, combatants engage each other with rifles, bayonets, sticks, or entrenching tools. See figure below.

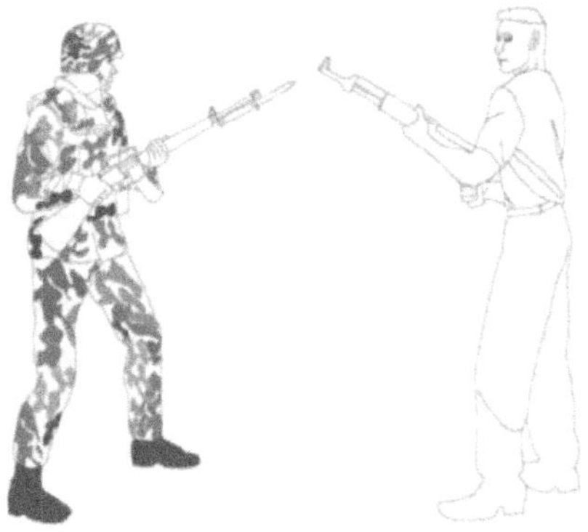

Midrange

During midrange engagements, combatants engage each other with knives, punches, or kicks.

Close Range

During close range engagements, combatants grab each other. Close range engagements also involve elbow strikes, knee strikes, and grappling.

2. Weapons of the Body

Hands and Arms

The hands, forearms, and elbows are the arm's individual weapons. The hands consist of several areas that can be used as weapons: fists, edges of hands, palms, and fingers.

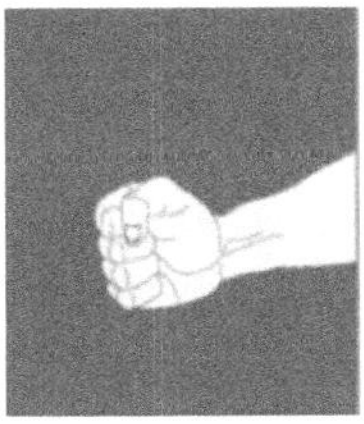

Fists. To minimize injury to the fists, Marines use their fists as weapons to target soft tissue areas such as the throat. The fists' striking surfaces are the first two knuckles of the hands or the meaty portions of the hands below the little fingers.

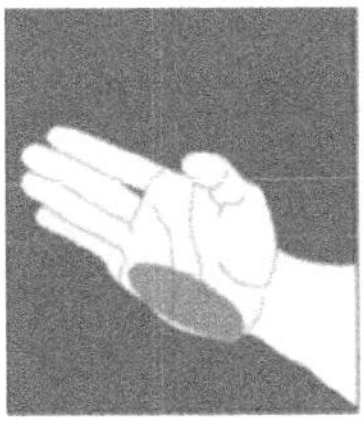

Edge of Hand. Marines use the edge of the hand (knife edge) as a weapon. Marines use the edge of the hand to strike soft tissue areas.

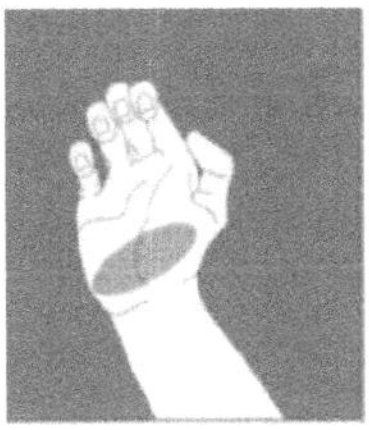

Palms. Because of the palm's padding, Marines use the heels of the palms to strike, parry, and/or block.

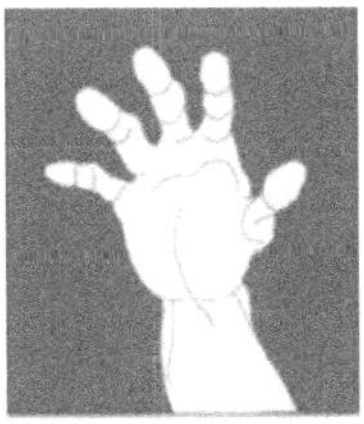

Fingers. Marines use the fingers to gouge, rip, and tear soft tissue areas (e.g., eyes, throat, groin).

Forearms. Marines use the forearms as a defensive tool to deflect or block attacks. Forearms can also be used as striking weapons to damage or break an opponent's joints and limbs. Marines sustain less self-injury when strikes are conducted with the forearms than when strikes are conducted with fists and fingers.

Elbows. Marines use the elbows as striking weapons. Because of the short distance needed to generate power, elbows are excellent weapons for striking during the close range of close combat.

Legs

The legs are more powerful than any other weapon of the body, and they are less prone to injury when striking. The feet are protected by boots and are the preferred choice for striking.

Feet. Marines use the balls of the feet, the insteps, and the toes to kick an opponent. Marines use the cutting edge of the heels and the heels to stomp on an opponent. Marines must be wearing boots when striking with the toes.

Knees. Like elbows, knees are excellent weapons in the close range of close combat. Knee strikes are most effective while fighting close to an opponent where kicks arc impractical. The opponent's groin area is an ideal target for the knee strike if he is standing upright. Knee strikes can deliver a devastating secondary attack to an opponent's face following an initial attack that caused him to bend at the waist.

3. Target Areas of the Body

During close combat, Marines strive to attack the accessible target areas of an opponent's body. The readily accessible areas will vary with each situation and throughout the engagement. The target areas are divided into five major groups: head, neck, torso, groin, and extremities. The figure below illustrates target areas of the body.

Head

The vulnerable regions of the head are the eyes, temple, nose, ears, and jaw. Massive damage to the head kills an opponent.

Eyes. The eyes are excellent targets because they are soft tissue areas that arc not protected by bone or muscle. Attacks to this area may cause the opponent to protect the area with his hands, allowing Marines to execute a secondary attack to other

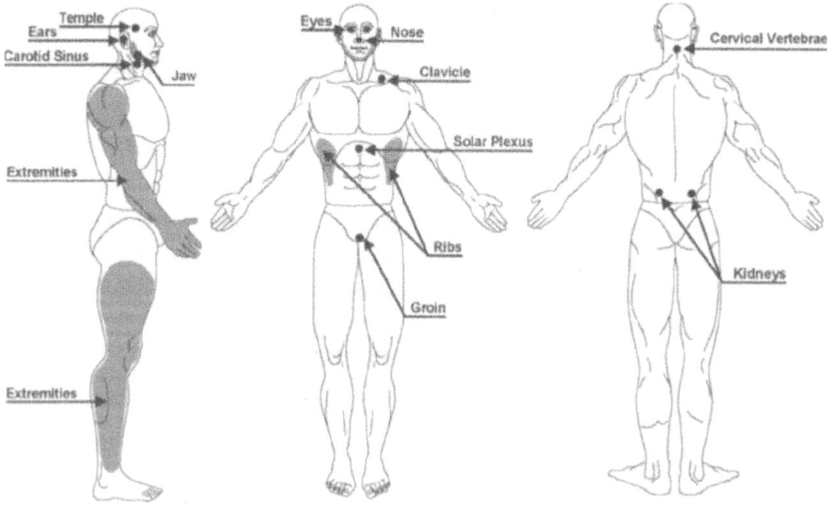

target areas while the opponent uses his hands to protect his eyes.

Temple. The temple is one of the most fragile areas of the head. Powerful strikes to the opponent's temple cause permanent damage and death.

Nose. The nose is very sensitive and easily broken. An attack to this area causes involuntary watering and closing of the opponent's eyes, rendering him vulnerable to secondary attacks. However, through training, individuals can condition themselves to withstand attacks to the nose. Therefore, any attack to the nose must be powerfully delivered.

Ears. Attacks to the ears may cause the eardrum to rupture. But this may not stop or even distract an opponent unless Marines powerfully deliver the strike.

Jaw. The jaw region, when struck forcefully, renders the opponent unconscious. Strikes to the jaw cause painful injuries to the teeth and surrounding tissues (e.g., lips, tongue), but the risk of self-injury is great unless Marines deliver strikes with a hard object such as a helmet, rifle butt, or boot heel.

Neck

The front of the neck, or throat area, is a soft tissue area that is not covered by natural protection. Damage to this region causes the opponent's trachea to swell, closing his airway, which can lead to death.

Carotid Sinus. The carotid sinus is located on both sides of the neck just below the jaw. Strikes to the carotid sinus restrict blood flow to the brain, causing loss of consciousness or death.

Cervical Vertebrae. The cervical vertebrae on the back of the neck, from the base of the skull to the top of the shoulders, contains the spinal cord, which is the nervous system's link to the brain. The weight of the head and the lack of large muscle mass allow damage to the cervical vertebrae and spinal cord. Excessive damage to this area causes pain, paralysis, or death.

Torso

Clavicle. The opponent's clavicle (or collar bone) can be easily fractured, causing immobilization of the arm.

Solar Plexus. Attacks to the opponent's solar plexus or center of the chest can knock the breath out of him and immobilize him.

Ribs. Damage to the opponent's ribs immobilizes him. It may also cause internal trauma.

Kidneys. Powerful attacks to the opponent's kidneys cause immobilization, permanent damage, or death.

Groin

The groin area is another soft tissue area not covered by natural protection. Any damage to this area causes the opponent to involuntarily protect his injured area, usually with his hands or legs. In male opponents, the scrotum is the main target since even a near miss causes severe pain, contraction of the lower abdominal muscles, deterioration of his stance, and possible internal trauma.

Extremities

Rarely will an attack to the opponent's extremities (arms and legs) cause death, but they are important target areas in close combat. Damage to an opponent's joints causes immobilization.

4. Pressure Points of the Body

There are nerves in the human body that, when pressure is applied or when they are struck, allow Marines to control a subject through pain compliance. Marines use pressure points to control an opponent when deadly force is not authorized. They also use pressure points to soften or distract an opponent so a lethal or nonlethal technique can be employed. The figure on page 1-5 illustrates the body's pressure points. Marines execute attacks to pressure points by—

: Rapidly kicking or striking pressure points.
: Slowly applying steady pressure to pressure points.

Infraorbital Nerve

The infraorbital nerve is just below the nose. Marines apply pressure to this nerve with an index finger to control the opponent.

Mastoid Process

The mastoid process is behind the base of the ear and beneath the edge of the jaw. Marines apply inward and upward pressure to this pressure point with the fingers to distract and control the opponent.

Jugular Notch

The jugular notch is at the base of the neck in the notch formed at the center of the clavicle. Marines apply pressure in a quick, stabbing motion with the index finger. Strikes to the jugular notch cause serious damage.

Brachial Plexus (Tie In)

The brachial plexus (tie in) is on the front of the shoulder at the joint. Strikes and pressure applied with the hand are effective on this nerve.

Radial Nerves

Radial nerves are on the inside of the forearms along the radius bones. Strikes and pressure applied with the hand to the radial nerve serve as a softening technique.

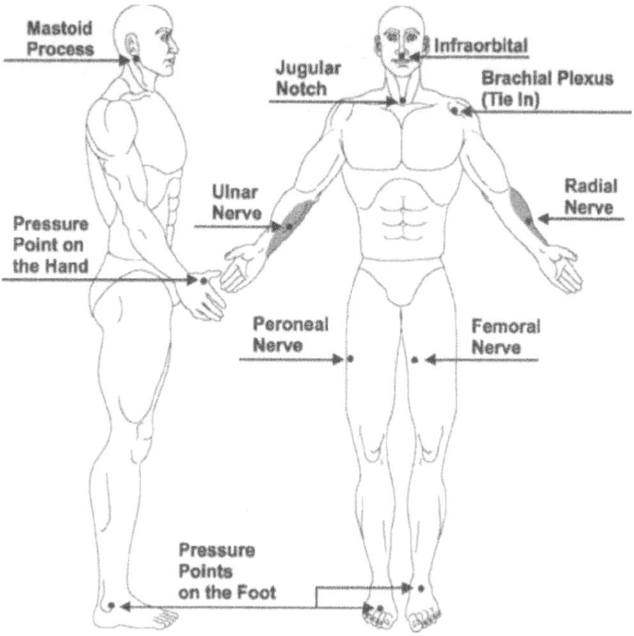

Ulnar Nerve

Ulnar nerves are on the outside of the forearms along the ulnar bones. Strikes and pressure applied with the hand to the ulnar nerve serve as a softening technique.

Pressure Point on the Hand

The hands contain a pressure point on the webbing between the thumbs and index fingers where the two bones of the fingers meet.

To force an opponent to soften or release his grip, Marines apply pressure with their index fingers to this pressure point or strike this pressure point with their fists.

Femoral Nerves

Femoral nerves are on the inside of the thighs along the femur bones. Strikes to the femoral nerve serve as a softening technique.

Peroneal Nerves

Peroneal nerves are on the outside of the thighs along the femur bones. Strikes to the peroneal nerve serve as a softening technique.

Pressure Points on the Feet

There are pressure points on the feet that, when pressure is applied or when they are struck, serve to soften or distract the opponent. Marines apply pressure with the toe, edge, or heel of their boots to the following points:

: The notch below the ball of the ankle.
: The top center of the foot, above the toes.
: The top of the foot where the leg and foot meet.

5. Basic Warrior Stance

Achieving a solid stance is critical to stability and movement throughout any close combat confrontation. The basic warrior stance provides the foundation for stability and movement that is needed to execute close combat techniques. To execute the basic warrior stance, Marines put their feet apart, hands up, elbows in, and chin down.

Feet Apart

Place feet shoulder-width apart.

Keep the head forward and eyes on the opponent, take a half step forward with the left foot, and pivot on the heels so the hips and shoulders are at approximately a 45-degree angle to the right.

Distribute body weight evenly on both legs. Bend the knees slightly.

Hands Up

Curl the fingers naturally into the palm of the hand. Position the thumb across the index and middle fingers. Do not clench the fists. Clenching the fists increases muscular tension in the forearms and decreases speed and reaction time.

Bring the hands up to the face at chin level, with the palms facing each other. Hold the fists up high enough to protect the head, but not so high that they block the field of vision. Ensure continuous eye contact with the opponent.

Elbows In

Tuck the elbows in close to protect the body.

Chin Down

Tuck the chin down to take advantage of the natural protection provided by the shoulders.

6. Angles of Approach and Movement

Marines use movement to control a confrontation and to retain a tactical advantage. Movement increases power and maximizes momentum. By moving around the opponent, Marines gain access to different target areas of the opponent's body. Movement allows Marines to use different weapons of the body and different close combat techniques to attack specific target areas.

Angles of Approach

Marines move anywhere within a 360-degree circle around the opponent to gain a tactical advantage. This circle provides access to different target areas of the opponent's body.

When facing an opponent, Marines move in a 45-degree angle to either side of the opponent. Moving at a 45-degree angle avoids an opponent's strike and puts Marines in the best position to attack the opponent. Marines should avoid being directly in front of an opponent during a confrontation. If a Marine is directly in front of an opponent, the opponent can rely on his forward momentum and linear power to seize the tactical advantage.

Movement

Marines must know how to move in all directions while maintaining the basic warrior stance. During any movement, the legs or feet should not be crossed. Once a movement is completed, the basic warrior stance should be resumed. Maintaining the basic warrior stance protects Marines and puts them in the proper position to launch an attack against an opponent.

Note: Before body movement begins. Marines turn their heads quickly to the new direction. The faster the head turns, the faster the body moves, and the quicker Marines attain visual contact with the opponent.

Forward to the Left. To move forward to the left from the basic warrior stance, Marines-

: Move the left foot forward at a 45-degree angle from the body (approximately 12 to 15 inches), keeping the toe pointed toward the opponent.

Bring the right foot behind the left foot as soon as the left foot is in place. This returns Marines to the basic warrior stance.

Forward to the Right. To move forward to the right from the basic warrior stance, Marines-

Move the right foot forward at a 45-degree angle from the body (approximately 12 to 15 inches).

Bring the left foot, toe pointing toward the opponent, in front of the right foot as soon as the right foot is in place. This returns Marines to the basic warrior stance.

Backward to the Left. To move backward to the left from the basic warrior stance, Marines execute the forward movement in reverse. Marines-

Move the left foot backward at a 45-degree angle from the body (approximately 12 to 15 inches), keeping the toe pointed toward the opponent.

Bring the right foot behind the left foot as soon as the left foot is in place. This returns Marines to the basic warrior stance.

Backward to the Right. To move backward to the right from the basic warrior stance, Marines execute the forward movement in reverse. Marines-

Move the right foot backward at a 45-degree angle from the body (approximately 12 to 15 inches).

Bring the left foot, toe pointing toward the opponent, in front of the right foot as soon as the right foot is in place. This returns Marines to the basic warrior stance.

7. Balance and Off-Balancing

Balance

In any close combat situation, Marines must strive to maintain balance. The last place to be in a close combat situation is on the ground. Marines must maintain a strong base and a low center of balance, their feet must be a shoulder-width apart, and they must stay on their toes to enable quick movement.

Off-balancing

Marines use off-balancing techniques to control an opponent. These techniques are used to throw an opponent to the ground while Marines remain standing, or they are used to put Marines in a position for an offensive attack.

Off-balancing techniques use the opponent's momentum to move or throw him. For example, if the opponent is charging a Marine, the Marine pulls the opponent to drive him to the ground. Likewise, if the opponent is pulling a Marine, the Marine pushes the opponent to drive him to the ground.

Off-balancing techniques also rely on the power generated by the opponent. For example, during combat a Marine may be tired or outnumbered. Depending on the generated energy and momentum of the opponent, the Marine employs off-balancing techniques with very little effort and still provides effective results.

Because off-balancing techniques rely on the momentum and power generated by the opponent, these techniques are particularly effective for Marines who may be outsized by their opponent or lack their opponent's strength.

Angles of Off-balancing

There are eight angles or directions in which an opponent can be off-balanced: forward, rear, right, left, forward right, forward left, rear right, and rear left.

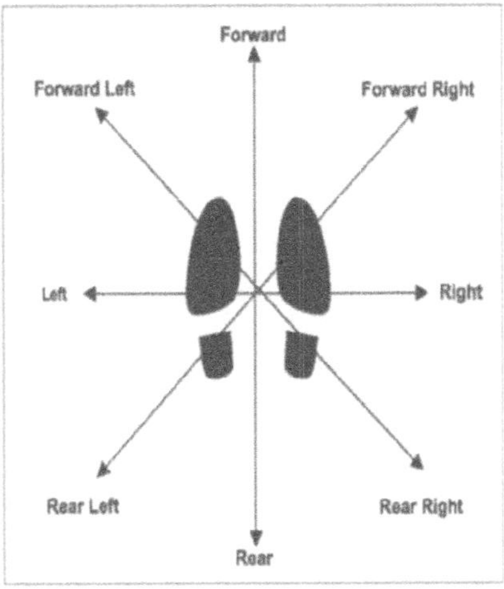

Note: The angles correspond to the Marine's perspective, not the opponent's. Forward, rear, right, and left are straight angles. Forward right, forward left, rear right, and rear left are considered quadrants that are at a 45-degree angle in either direction to the front or rear.

Off-balancing Techniques

Marines off-balance an opponent by pushing, pulling, or bumping the opponent with their hands, arms, or bodies.

To pull, Marines grab an opponent with their hands and drive him forcefully to one of the rear quadrants or to the right or left.

To push, Marines grab the opponent with their hands and drive him forcefully into one of the front quadrants or to the right or left.

Marines execute bumping in the same manner as pushing, but use their shoulders, hips, and legs instead of their hands to off-balance the opponent.

8. Falls

Marines may lose their balance or be thrown to the ground during encounters with an opponent. Marines use falling techniques to absorb the impact of a fall and to quickly return to their feet following an opponent's attack.

Whether falling or being thrown by an opponent, Marines strive to reduce the force of the impact, to prevent serious personal injury, and to increase the chances of survival. Falling techniques use the body's large muscles (back, thighs, buttocks) to protect vital organs and bones from injury and immobilization.

Front Fall

Marines execute a front fall to break a fall on the front. To execute the front fall, Marines-

Bend the elbows and place the palms facing out in a position to spread and absorb the impact of the fall.

Fall forward, breaking the fall with the forearms and palms. The forearms and hands, down to the fingertips, should strike the ground simultaneously.

Offer resistance with the forearms and hands to keep the head raised off the ground.

Side Fall

Marines execute a side fall to break a fall on the side. To execute the side fall, Marines-

Bring the right arm across the body so the hand is next to the left shoulder with the palm facing inboard.

Fall to the side, breaking the fall with the right arm by slapping the ground and making contact from the shoulder or forearm down to the hand. At the same time, tuck the chin and keep the head raised off the ground. The chin should be tucked to the chest at all times to prevent whiplash.

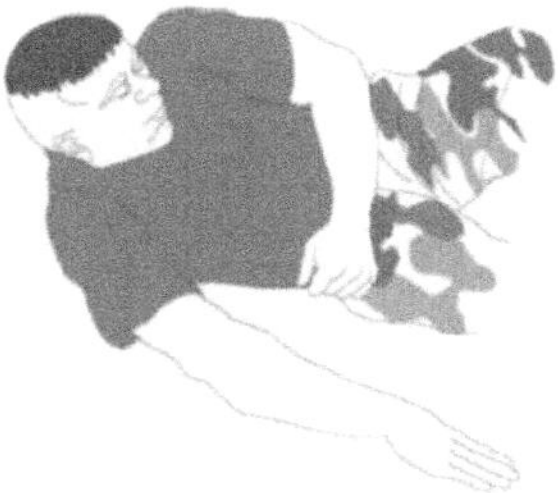

Stretch the right leg out to make contact with the ground and to distribute and absorb the impact. Bend the left leg, allowing the foot to make contact with the ground.

Back Fall

Marines execute a back fall to break the fall when being thrown or falling backward. To execute the backfall, Marines-

Cross the hands in front of the chest and tuck the chin.

Fall backward and slap the ground with the forearms and hands to absorb the impact of the fall and keep the head off the ground.

Forward Shoulder Roll

Marines use the forward shoulder roll to break a fall from an opponent's attack and to use the momentum of the fall to get back on their feet quickly. Ideally, Marines execute the forward shoulder roll to a standing position so they can continue fighting. To execute the forward shoulder roll, Marines-

Contact the ground with the back of the right forearm and upper arm. Tuck the chin into the chest.

Roll onto the right shoulder, rolling diagonally across the back to land on the left hip.

Slap the ground with the left arm, absorbing the impact from the shoulder to the hand, palm down.

Keep the left leg straight to absorb as much of the impact as possible. The right leg is bent and the foot hits flat on the ground.

Bend the left leg upon impact to push off with the left knee and leg to a squatting and then a standing position. Forward momentum should carry the Marine to a standing position.

CHAPTER 2

LETHAL AND NONLETHAL WEAPONS TECHNIQUES

This chapter describes all techniques for a right-handed person. However, all techniques can be executed from either side.

In drawings, the Marine is depicted in woodland camouflage utilities; the opponent is depicted without camouflage. In photographs, the Marine is depicted in woodland camouflage utilities; the opponent is depicted in desert camouflage utilities.

1. Bayonet Techniques

WARNING

During training, Marines must have bayonets sheathed. Marines use bayonet dummies to practice bayonet techniques. When practicing offensive and defensive bayonet techniques student-on-student, Marines use pugil sticks.

All Marines armed with a rifle carry a bayonet. The bayonet is an effective weapon if Marines are properly trained in offensive and defensive bayonet techniques. An offensive attack, such as a thrust, is a devastating attack that can quickly end a fight. Defensive techniques, such as the block and parry, can deter the opponent's attack and allow Marines to regain the initiative. Through proper training, Marines develop the courage and confidence required to effectively use a bayonet to protect themselves and destroy the enemy. In situations where friendly and enemy troops are closely mingled and rifle fire and grenades are impractical, the bayonet becomes the weapon of choice.

Holding the Rifle

To execute bayonet techniques, Marines hold the rifle in a modified basic warrior stance. All movement begins and ends with the basic warrior stance. To hold the rifle, Marines—

Use an overhanded grasp to grab the small of the rifle's stock. Use an underhanded grasp to grab the handguards of the rifle.

Lock the buttstock of the rifle against the hip with the right forearm.

Orient the blade end of the rifle toward the opponent.

Offensive Bayonet Techniques

Straight Thrust. Marines use the straight thrust to disable or kill an opponent. It is the most deadly offensive technique because it causes the most trauma to an opponent. Target areas are the opponent' s throat, groin, or face. The opponent's chest and stomach are also excellent target areas if not protected by body armor or combat equipment. To execute the straight thrust, Marines—

Lift the left leg and lunge forward off the ball of the right foot while thrusting the blade end of the weapon forward, directly toward the opponent.

Retract the weapon and return to the basic warrior stance.

Slash. Marines use the slash to kill an opponent or to create an opening in his defense. The target area is the opponent's neck. To execute the slash, Marines—

Extend the left hand back toward the left shoulder.

Thrust the left hand forward and swing it to the right, bring the right hand back toward the hip, and turn the cutting edge of the blade toward the opponent's neck. The movement is a slashing motion so the blade cuts across the opponent's neck.

Horizontal Buttstroke. Marines use the horizontal buttstroke to weaken an opponent's defenses, to cause serious injury, or to set him up for a killing blow. Target areas are the opponent's head, neck, and legs. To execute the horizontal buttstroke, Marines—

Step forward with the right foot and drive the right hand forward. Rotate the hips and shoulders into the strike. Move the left hand back toward the left shoulder.

Strike the opponent with the butt of the weapon.

Vertical Buttstroke. Marines use the vertical buttstroke to weaken an opponent's defenses, to cause serious injury, or to set him up for a killing blow. Target areas are the opponent's groin and face. To execute the vertical buttstroke, Marines—

Step forward with the right foot and drive the right hand straight up.

Pull the left hand back over the left shoulder.

Strike the opponent with the butt of the weapon.

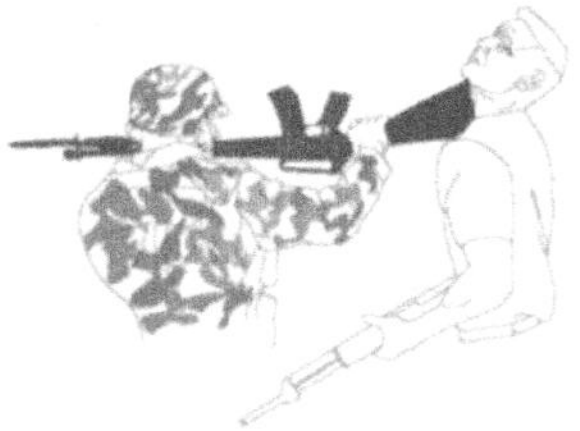

Smash. Marines use the smash as a follow-up technique to the horizontal or vertical buttstroke, particularly if they missed the target. The target area is the opponent's head. To execute the smash following a buttstroke, Marines-

Step forward with the right foot and place the blade end of the weapon over the left shoulder and elevate the right elbow above the shoulders.

Drive the arms straight forward, striking the opponent with the butt of the weapon.

Defensive Bayonet Techniques

Parry. Marines use a parry as a defensive technique to redirect or deflect an attack. A parry is a slight redirection of a linear attack by an opponent; e.g., a straight thrust or a smash. To execute the parry, Marines-

Use the bayonet end of the rifle to redirect the barrel or bayonet of the opponent's weapon.

Lock the weapon against the hip with the right forearm.

Rotate to the right or left, moving the bayonet end of the rifle to parry the opponent's attack. Rotation is generated from the hips.

Redirect or guide the opponent's weapon away from the body by exerting pressure against the opponent' s weapon.

High Block. Marines execute a high block against a vertical attack coming from high to low. To execute the high block, Marines—

Thrust the arms up forcefully at approximately a 45-degree angle from the body. The weapon should be over the top of the head and parallel to the ground. The elbows are bent, but there is enough muscular tension in the arms to absorb the impact and deter the attack.

Low Block. Marines execute the low block

against a vertical attack coming from low to high. To execute the low block, Marines—

Thrust the arms down forcefully at approximately a 45-dcgrcc angle from the body. The weapon

should be at or below the waist and parallel to the ground. The elbows are bent, but there is enough muscular tension in the arms to absorb the impact and deter the attack.

Left and Right Block. Marines execute a left or right block against a horizontal buttstroke or a slash. To execute the left or right block, Marines-

Thrust the arms forcefully to the right or left, holding the rifle vertically in the direction of the attack. The elbows are bent, but there is enough muscular tension in the arms to absorb the impact and deter the attack.

Counter Action Following the Block. After deflecting an opponent's attack with a block, Marines counter with a slash or a horizontal buttstroke to regain the initiative. However, the objective in any bayonet fight is to thrust forward with the blade end of the weapon to immediately end the fight.

Group Strategy

On occasion, Marines may engage an opponent as a member of a group or numerous opponents by one's self or as a member of a group. By combining bayonet fighting movements and simple strategies, Marines can effectively overcome their opponent or opponents

Offensive Strategy: Two Against One. If two bayonet fighters engage one opponent, the fighters advance together.

Fighter 1 engages the opponent while fighter 2 swiftly and aggressively attacks the opponent's exposed flank and destroys the opponent.

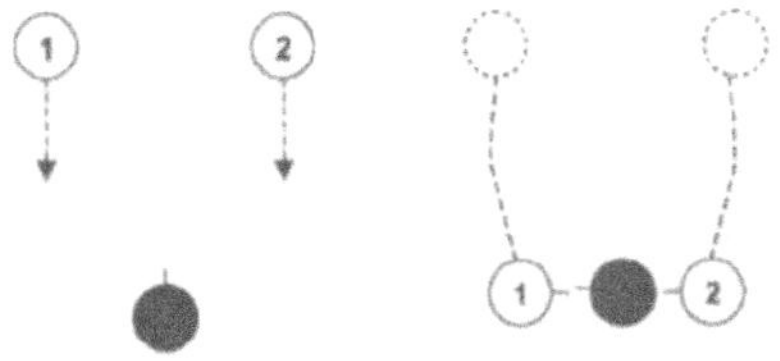

Offensive Strategy: Three Against Two. If three bayonet fighters engage two opponents, the fighters advance together keeping their opponents to the inside.

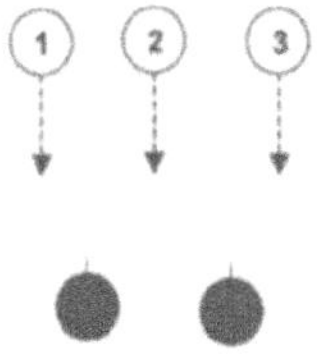

Fighters 1 and 3 engage opponents. Fighter 2 attacks the opponent's exposed flank engaged by fighter 1 and destroys the opponent.

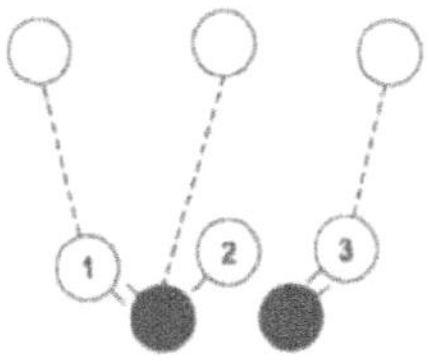

Fighters 1 and 2 turn and attack the exposed flank of the opponent engaged by fighter 3 and destroy the opponent.

Defensive Strategy: One Against Two. If a fighter is attacked by two opponents, the fighter immediately positions himself at the flank of the nearest opponent and keeps that opponent between himself and the other opponent.

Using the first opponent's body as a shield against the second opponent, the fighter destroys the first opponent quickly before the second opponent moves to assist.

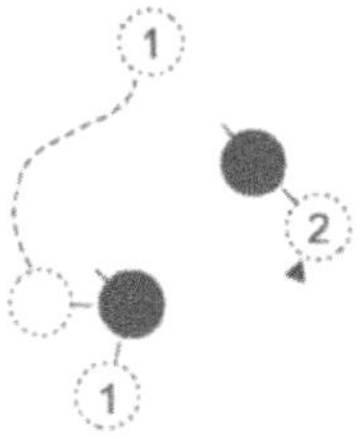

Then, the fighter engages and destroys the second opponent.

Defensive Strategy: Two Against Three. If two fighters arc attacked by three opponents, the fighters immediately move to the opponent's flanks.

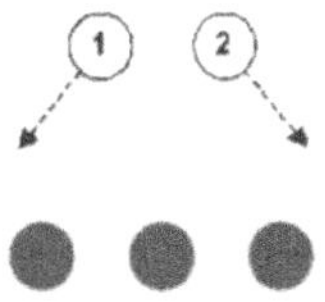

Fighters 1 and 2 quickly attack and destroy their opponents before the third opponent closes in.

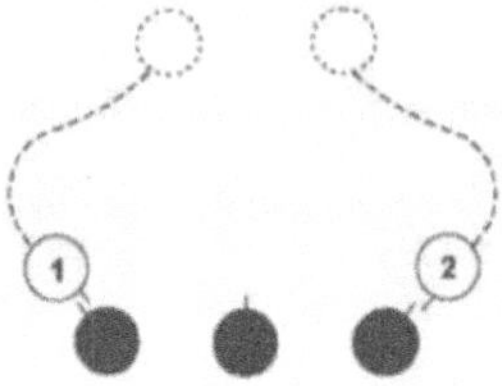

Fighter 1 engages the third opponent while fighter 2 attacks the opponent's exposed flank and destroys the opponent.

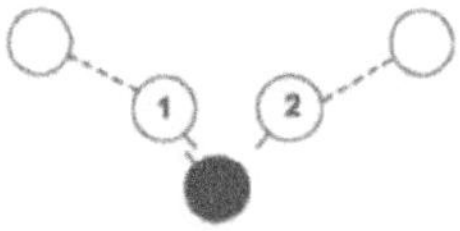

2. Nonlethal Rifle and Shotgun Retention Techniques

Most Marines are armed with the M16A2 service rifle. Marines are taught to keep their weapon with them at all times. Marines must be constantly alert to their surroundings and the people moving in and around their environment. Marines may be confronted by an individual who tries to take theii weapons. If this happens, Marines should not struggle with the individual. To retain positive control of their weapons, Marines must understand and apply weapons retention techniques, otherwise known as armed manipulation. The following techniques can be used with either the rifle or the shotgun.

Blocking Technique

To execute a blocking technique, Marines-

Stand in a defensive posture.

Use the weapon to block the opponent by thrusting it out firmly, with the elbows bent. Do not try to hit the opponent with the rifle. The rifle is used as a barrier.

Technique if Opponent Grabs Weapon Underhanded

Usually an opponent will try to grab the weapon or block it as an instinctive action. If the opponent uses an underhand grab to seize the handguards of the weapon, Marines —

Trap the opponent's closest finger(s) above the knuckle with the thumb so he cannot release his grip-

Apply bone pressure on the opponent's finger to initiate pain compliance.

Rotate the barrel of the weapon up or down quickly while maintaining pressure on the opponent's hand. At the same time, quickly pivot to off-balance the opponent.

Technique if Opponent Grabs Weapon Overhanded

If the opponent uses an overhand grab to seize the weapon overhanded, Marines—

Trap the opponent's finger to hold his hand in place.

Rotate the barrel to place it across the opponent's forearm and apply downward pressure. This action is similar to an armbar.

Technique if the Opponent Grabs the Muzzle

If the opponent grabs the muzzle of the weapon, Marines—

Rotate the muzzle quickly in a circle motion, Slash downward with the muzzle to release the opponent's grip.

Butt Strikes

Strikes with the butt of the weapon control or ward off an attacker. During any of the retention techniques, Marines use the heel or cutting edge of the weapon to deliver butt strikes to the inside or the outside of the opponent's thigh. The inside butt strike targets the femoral nerve as illustrated in the photo to the left.

The outside butt strike targets the peroneal nerve as illustrated in the photo to the left. Strikes can be made to the outside or inside of the thighs. If a strike to one side of the thigh misses, Marines follow back through with the butt of the weapon on the other side of the thigh.

Off-balancing Techniques

Marines apply off-balancing techniques to throw an opponent to the ground and retain possession of the weapon.

If the opponent grabs the weapon and pushes, Marines should not push on the weapon. They should—

Move with the momentum and movement of the

opponent by pivoting in the direction of the movement by stepping back.

Throw the opponent to the ground with a quick jerking movement by lowering the mu/./le and swinging the butt of the weapon.

If the opponent grabs the weapon and pulls, Marines—

Step on the opponent's foot and push forward to off-balance him and drive him to the ground.

Sweep the opponent's feet out from under him by hooking his leg with the leg and kicking backward.

3. Nonlethal Handgun Retention Techniques

Many Marines are armed with the M9 service pistol. Marines must keep their weapons in their possession at all times. Marines must be constantly alert to their surroundings and the people moving in and around the environment. Marines may be confronted by an individual who tries to take their weapons. To retain positive control of the weapon, Marines must understand and apply handgun retention techniques.

Blocking Technique

Marines perform the following blocking techniques if an opponent attempts to grab their pistol in the holster. Marines—

Place the body between the weapon and the opponent by immediately pivoting so the weapon is away from the opponent.

Step back and away from the opponent while placing the hand on the pistol grip.

Extend the left hand and block, deflect, or strike the opponent's arm.

Armbar Technique

Marines use the armbar technique when an opponent uses his right hand to grab their pistol in the holster. To execute the armbar technique, Marines—

Trap the opponent's right hand by grasping the opponent's wrist or hand with the right hand and applying pressure against the body.

Step back with the right foot and pivot sharply to the right to be next to the opponent. Always pivot in a direction that keeps the weapon away from the opponent.

Straighten the opponent's arm to apply an armbar. The arm should be straight across the torso.

Continue pivoting to the right while pulling back on the opponent's shoulder. This action may break the opponent's arm.

Place the left hand across the opponent's face and apply pressure back and down to take the oppo-

nent to the ground. Pressure applied beneath the nose or on the trachea is equally effective.

Wristlock Technique

Marines use the wristlock technique when an opponent grabs their pistol while it is in the holster with his right hand. To execute the w ristlock technique, Marines—

Grasp the opponent's wrist or hand with the right hand and apply pressure against the body.

Step back and away from the opponent and pivot to the right so the weapon is away from the opponent. Always pivot in a direction that keeps the weapon away from the opponent.

Reach over the opponent's arm with the left hand and grab his hand, applying pressure against his arm with the left forearm. Execute a wristlock.

Incorporate the second hand into the wristlock and, stepping back with the left foot, pivot to the left.

Execute a two-handed wristlock by exerting downward pressure with the thumb and rotating his hand to the left.

Continue to pivot to off-balance the opponent and drive him to the ground.

Note: If the opponent grabs the pistol with his left hand, Marines execute the wristlock with one hand and step in toward the opponent, rather than

away from the opponent.

Softening Techniques

Handgun retention techniques use softening techniques applied to pressure points. Bone pressure

and strikes with the hands (i.e., hammer fist), knees, and feet are also effective softening techniques.

Pressure Points. Marines use pressure point techniques to get the opponent to loosen his grip. Marines use their finger tips to apply pressure to the webbing between the index finger and thumb, the jugular notch, and the brachial plexus tie in. The following figure illustrates pressure applied to the brachial plexus tie in.

Bone Pressure. Bone pressure is the application of pressure on a bone against a hard object to initiate pain compliance. To apply bone pressure, Marines use their hand to trap the opponent's hand on the weapon. Marines apply a slow, steady pressure to the opponent's hand and fingers until his grip is softened or he releases his hold.

Strikes. If it is difficult to apply a retention technique, Marines employ strikes or kicks to force the opponent to loosen his grip. Strikes to the

eyes, the arms (radial nerve), or shoulder (brachial plexus tic in) soften the opponent's grip on the weapon.

Kicks and knee strikes to the peroneal nerve, the femoral nerve, or the groin arc effective because the opponent is typically unprepared to counter the strike.

Stomping on the top of the opponent's foot may distract him or loosen his grip on the weapon.

4. Firearm Disarmament Techniques

Marines use firearm disarmament techniques during a close-range confrontation if they are un-

armed and the opponent has a firearm (pistol). These techniques are equally effective if Marines are armed but do not have time to withdraw and present the weapon. The goal of firearm disarmament techniques is to gain control of the situation

so Marines gain the tactical advantage. The goal is not necessarily to get control of the opponent's weapon.

Pistol to the Front

This technique is used when Marines are unarmed and the opponent has a pistol pointing at their front (e.g., head, chest). The technique is the same if the opponent sticks the pistol under the Marine's chin. To execute the counter when an opponent is pointing a pistol toward the front of a Marine, Marines—

Place the hands close to the weapon, about chest high, palms out.

Use the left hand to grab the opponent's forearm and push the opponent's hand with the pistol away to clear the body from in front of the weapon. At the same time, rotate the right shoulder back to clear the body from the weapon.

Maintain control of the opponent's arm.

Grasp the weapon with the right hand by placing the thumb underneath the pistol and the fingers over top of the pistol.

Keep the right hand wrapped tightly around the muzzle and quickly rotate the pistol in the opponent's hand so the muzzle is facing the opponent.

Grasp and pull the opponent's wrist or forearm away from the body while rotating the weapon.

Rotate the weapon toward the opponent while pulling it up and back and out of the opponent's grasp.

Pistol to the Rear

This technique is performed when Marines are unarmed and the opponent has a pistol pointing to the back of the Marine's head. To execute the counter when the opponent is pointing a pistol to the rear, Marines—

Place the hands close together about chest high, palms out.

Step back with the left foot, pivoting on the right foot so the side is against the opponent's front. This action clears the body from the weapon's line of fire. Keep the left hand up.

Pivot on the left foot to face the opponent and, at the same time, raise the left elbow and reach over the top of the opponent's arm with the left arm.

Wrap the left arm tightly around the opponent's arm above his elbow to control it.

Push on the opponent's shoulder with the right hand while pulling up with the left arm to achieve an armbar. This action releases the opponent's grip on the weapon. If necessary, the opponent can be taken to the ground with a leg sweep.

Note: To execute this technique, the weapon must be close to or touching the rear of the Marine. If the weapon is too far away from the body, this technique would be difficult to execute or it would be ineffective.

CHAPTER 3

HAND-HELD WEAPONS

This chapter describes all techniques for a right-handed person. However, all techniques can be executed from either side.

In drawings, the Marine is depicted in woodland camouflage utilities; the opponent is depicted without camouflage. In photographs, the Marine is depicted in woodland camouflage utilities; the opponent is depicted in desert camouflage utilities.

Marines must know how to defend against attacks when an opponent is either unarmed or armed with a held-held weapon. This chapter addresses the combative use of knives, specific weapons of opportunity, and sticks. However, virtually anything can be used as a hand-held weapon.

1. Fundamentals of Knife Fighting

Marines must be trained in knife fighting techniques. Marines experienced in offensive knife techniques can cause enough damage and massive trauma to stop an opponent. When engaged against each other, experienced knife fighters employ various maneuvers and techniques that are specific to knife fighting. Seldom, if ever, will Marines engage an opponent in a classical knife fight.

Note: When armed with a rifle, Marines are issued a bayonet. When armed with a pistol, Marine are issued a combat knife.

Angles of Attack

There are six angles from which an attack with a knife can be launched:

- Vertical strike coming straight down on an opponent.
- Forward diagonal strike coming in at a 45-degree angle to the opponent.
- Reverse diagonal strike coming in at a 45-degree angle to the opponent.
- Forward horizontal strike coming in parallel to the ground.
- Reverse horizontal strike coming in parallel to the ground.
- Forward thrust coming in a straight line to the opponent.

Target Areas of the Body

During any confrontation, the parts of the opponent's body that are exposed or readily accessible will vary. The goal in a knife fight is to attack the body's soft, vital target areas that are readily accessible (e.g, the face, the sides and front of the neck, the lower abdomen [or groin]).

Neck. Carotid arteries, located on either side of the neck, are good target areas because they are not covered by body armor or natural protection.

Lower Abdomen (or Groin). The lower abdomen (or groin region) is a good target area because it is not covered by body armor.

Heart. The heart, if not covered by body armor, is an excellent target which, if struck, can prove fatal in a matter of seconds or minutes.

Secondary Targets. There are secondary target areas that will cause substantial bleeding if an artery is severed. These target areas are not immediately fatal, but can become fatal if left unattended. Attacks to—

The legs can cause a great deal of trauma and prove fatal. For example, the femoral artery located on the inside of the thigh is a large artery which, if cut, will cause extensive blood loss.
The brachial artery, located between the biceps and triceps on the inside of the arm, can cause extensive bleeding and damage.
The arm's radial and ulnar nerves can cause extensive bleeding and damage.

Movement

Marines can move anywhere within a 360-degree circle around the opponent. This allows accessibility to different target areas of the opponent's body. Marines should avoid being directly in front of an opponent because the opponent can rely on his forward momentum to seize the tactical advantage. If Marines face an opponent, movement is made in a 45-degree angle to either side of the opponent. This angle avoids an opponent's strike and places Marines in the best position to attack an opponent.

Wearing the Combat Knife

Marines must wear the combat knife where it is easily accessible and where it can best be retained. It is recommended the combat knife be worn on the weak side hip, blade down, sharp edge facing forward. Marines can place it behind the magazine pouch where it is easily accessible to them, but not easily grabbed by an opponent.

Grip

The grip on the knife should be natural. Marines grasp the knife's grip with the fingers wrapped around the grip naturally as it is pulled out of its sheath. This is commonly known as a hammer grip. The blade end of the knife is always facing the opponent.

Stance

Marines use the basic warrior stance as the foundation for knife techniques. The left hand is a vertical shield that protects either the ribs or the head and neck. The right elbow is bent with the blade pointing forward toward the opponent' s head. This position serves as an index point, where all techniques are initiated.

Principles of Knife Fighting

The following are key principles of knife fighting:

Execute movements with the knife blade within a box, shoulder-width across from the neck down to the waistline. The opponent has a greater chance of blocking an attack if the blade is brought in a wide, sweeping movement to the opponent.
Close with the opponent, coming straight to the target.
Move with the knife in straight lines.
Point the knife's blade tip forward and toward the opponent.
Apply full body weight and power in each of the knife techniques. Full body weight should be put into the attack in the direction of the blade's movement (slash or thrust).
Apply constant forward pressure with the body and blade to keep the opponent off-balanced.

2. Knife Fighting Techniques

Slashing Techniques

Marines use slashing techniques to close with an enemy. Slashing techniques distract the opponent or damage the opponent so Marines can close in. Typically, Marines target the opponent's limbs, but any portion of the body that is presented can become a target.

Vertical Slash Technique. The vertical slash follows a vertical line straight down through the target. To execute the vertical slash, Marines—

Thrust the right hand out and bring the weapon straight down on the opponent, continuing to drag the knife down through the opponent's body.

Maintain contact on the opponent's body with the blade of the knife.

Forward Slash Technique. The forward slash follows a straight line in a forehand stroke, across the target areas of either the neck (high diagonal slash) or abdominal region (low horizontal slash). To execute the forward slash, Marines-
Extend the right hand while simultaneously rotating the palm up until the knife blade makes contact with the opponent.

Snap or rotate the wrist through the slashing motion to maximize blade's contact with the opponent.

Drag the knife across the opponent's body, from right to left, in a forehand stroke. The movement ends with the forearm against the body and the knife at the left hip with its blade oriented toward the opponent.

Reverse Slash Technique. The reverse slash is a follow-up technique to a forward attack. It allows Marines both a secondary attack and the ability to resume the basic warrior stance. The reverse slash follows a straight line in a backhand stroke, across the target areas of either the neck (high diagonal slash) or abdominal region (low horizontal slash). To execute the reverse slash, Marines—

Extend the right hand while simultaneously rotating the palm down until the knife blade makes contact with the opponent.

Snap or rotate the wrist through the slashing motion to maximize the blade's contact with the opponent.

Drag the knife across the opponent's body, from left to right, in a backhand stroke. Maintain contact on the opponent's body with the blade of the knife.

Thrusting Techniques

The primary objective of knife fighting is to insert the blade into an opponent to cause extensive damage and trauma. This is done with a thrusting technique. Thrusting techniques are more effective than slashing techniques because of the damage they can inflict. However, Marines use slashing techniques to close with the enemy so that they are closer to the opponent, which allows them to use the thrusting technique.

Vertical Thrust. The thrusting motion follows a vertical line straight up through the target (low into the abdomen region or high into the neck). To execute the vertical thrust, Marines—

Thrust the right hand toward the target, inserting the knife blade straight into the opponent.

Pull the knife out of the opponent.

Forward Thrust. The forward thrust follows a straight line straight into the opponent's neck (high thrust) or abdominal region (low thrust). To execute the forward thrust, Marines—

Thrust the right hand, palm down, toward the target, inserting the knife blade straight into the opponent.

Rotate the palm up once the knife is inserted to twist the blade.

Drop the right elbow and bring the knife to the opposite side of the opponent's body from where it was inserted. At the same time, rotate the hips and shoulders downward to bring body weight to bear on the attack.

Reverse Thrust. The reverse thrust is a follow-up technique to a forward attack. It allows Marines both a secondary attack and the ability to resume the basic warrior stance. The reverse thrust follows a horizontal line straight into the opponent's neck (high thrust) or abdominal region (low thrust). To execute the reverse thrust, Marines-

Bend the right arm, crossing the arm to the left side of the body.

Thrust the right hand, palm up, toward the target, inserting the knife blade straight into the opponent.

Rotate the palm down to twist the blade once the knife is inserted.

Bring the knife to the opposite side of the opponent's body from where it was inserted.

3. Weapons of Opportunity

During an unarmed close combat situation, Marines use their bodies as weapons, but they should be ready and able to use anything around them as a weapon. For example, Marines could throw sand or liquid in an opponent's eyes to temporarily impair his vision or smash the opponent's head with a rock or helmet. Marines must use whatever means are available and do whatever it takes to take control of the situation and to win, or they face the possibility of losing their lives. Some weapons of opportunity are discussed in the following subparagraphs.

Entrenching Tool

An entrenching tool (E-tool) is commonly carried by Marines. It can be an excellent weapon, especially when sharpened. Marines can use the E-tool to block, slash, and thrust at an opponent.

Tent Pole and Pins

Marines can use tent poles and pins to block, strike, or thrust at an opponent.

Web Belt

Marines can stretch a web belt between their hands to block attacks by an opponent.

Battlefield Debris

Marines can use debris on the battlefield (e.g., sticks, glass, a sharp piece of metal) to cut, slash, or stab an opponent. They can also use other types of debris such as shovel or ax handles, boards, metal pipes, or broken rifles to strike an opponent or apply a choke.

Helmet

A helmet can be used to strike an opponent on an unprotected area like the head and face. Grasp the rim of the helmet and thrust the arms forward, striking the opponent with the top of the helmet.

4. Fundamentals of Combative Stick

On the battlefield, Marines must be ready and able to use anything as a weapon. They must learn and be able to use techniques that can be employed with most weapons of opportunity. Among these techniques are combative stick techniques. Combative stick techniques can be used with a stick, a club, a broken rifle, an E-tool, or even a web belt.

Angles of Attack

There are six angles from which an attack with a hand-held weapon can be launched:

Vertical strike coming straight down on an opponent.
Forward diagonal strike coming in at a 45-degree angle to the opponent.
Reverse diagonal strike coming in at a 45-degree angle to the opponent.
Forward horizontal strike coming in parallel to the ground.
Reverse horizontal strike coming in parallel to the ground.
Forward thrust coming in a straight line to the opponent.

Grip

Grasp the stick about 2 inches from its base.

Stance

The basic warrior stance serves as the foundation for combative stick techniques. The left hand becomes a vertical shield that protects the ribs or the head and neck. Depending on how heavy the weapon is, it should be held at a level approximately shoulder height.

Movement

Movement during combative stick techniques is the same as it is for other close combat techniques. Marines can move anywhere within a 360-degree circle around the opponent. This allows accessibility to different target areas of the opponent's body and gains a tactical advantage. Marines should avoid being directly in front of an opponent because the opponent can rely on his forward momentum to seize the tactical advantage. If Marines face an opponent, movement is made in a 45-degree angle to either side of the opponent. This angle avoids an opponent's strike and places Marines in the best position to attack an opponent.

5. Combative Stick Techniques

Strikes

Strikes are intended to inflict as much damage on an opponent as possible. Striking techniques apply to a weapon of opportunity such as a stick, a tent pole, a club, a broken rifle, an E-tool, or a Pipe-

Vertical Strike. To execute the vertical strike, Marines—

Bend the right arm, extending the weapon over the back of the right shoulder.

Rotate the forearm straight down off the elbow to bring the weapon down on the opponent.

Rotate the hips and shoulders forcefully toward the opponent.

Follow through with the strike by allowing the weight of the weapon to go through the target area of the body.

Forward Strike. To execute a forward strike, Marines—

Step forward with the left foot in the direction of the strike.

Bend the right arm with the elbow extending out to the right and the weapon extended over the right shoulder.

Rotate the forearm to the right of the elbow to bring the weapon down onto the opponent. At the same time, forcefully rotate the hips and shoulders toward the opponent.

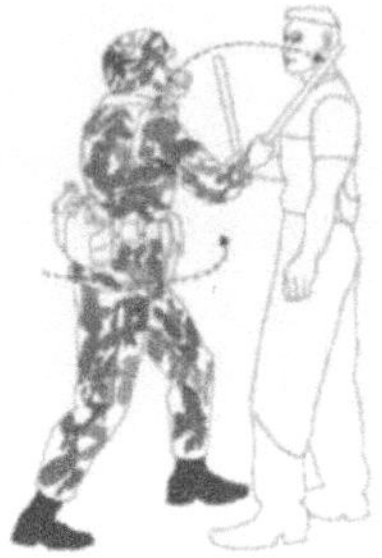

Follow through with the strike by allowing the weight of the weapon to go through the target area of the body.

Reverse Strike. The reverse strike is a follow-up technique to a forward strike. It allows Marines both a secondary attack and the ability to resume the basic warrior stance.To execute a reverse strike, Marines—

Step forward with the right foot in the direction of the strike.

Bend the right arm with the hand near the left shoulder. The weapon is extended over the left shoulder.

Rotate the forearm to the right of the elbow to bring the weapon down onto the opponent. At the same time, forcefully rotate the hips and shoulders toward the opponent.

Follow through with the strike by allowing the weight of the weapon to go through the target area of the body.

Forward Thrust. To execute the forward thrust, Marines—

Grasp the stick with the left hand, palm up, in a position where the stick can be controlled with two hands.

Lift the left leg and lunge forward off the ball of the right foot. At the same time, thrust the end of the weapon directly toward the opponent by thrusting both hands forward in a straight line.

6. Blocking Techniques

A block is meant to deter or deflect an attack by an opponent. A block sets up Marines for a follow-on attack against the opponent. Blocks are executed by deflecting, rather than hitting or following through like a strike.

Blocks Against Unarmed Attacks

To block against an unarmed attack, Marines—

Step forward at a 45-degree angle with either the right or left foot. This moves the body out of the line of attack.

Raise the left arm and block the strike with the meaty portion of the forearm.

Employ the stick by one of two techniques:

Strike the opponent with the stick.
Use the stick to block on two points of contact. When the stick is used to block, it serves as an incidental strike. This technique is only used if Marines have closed with the opponent and are inside his strike.

Blocks Against Armed Attacks

Block for a Vertical Strike. To execute the block for a vertical strike, Marines—

Step forward with the left foot at a 45-degree angle to the left. This moves the body out of the line of attack.

Block on two points of contact to disperse the impact of the attack:

Block the opponent's stick by positioning the stick so it is perpendicular to the opponent's stick. If the stick is not perpendicular to the opponent's stick, the stick can slide through and make contact on the Marine.
Block the opponent's wrist or forearm with the back of the left forearm.

Note: Use the stick to block the opponent's arm if closer to the opponent. It is the same movement as blocking with the arm, except the opponent's arm is blocked with both the stick and the arm.

Block for a Forward Strike. To execute the block for a forward strike, Marines-
Step forward with the left foot at a 45-degree angle to the left. This moves the body out of the line of attack and inside the opponent's strike.

Block on two points of contact to disperse the impact of the attack:

- Block the opponent's wrist or forearm with the meaty portion of the back of the left forearm.
- Strike the opponent's attacking biceps with the stick.

Block for a Reverse Strike. To execute the block for a reverse strike, Marines—

Step forward at a 45-degree angle with the right foot to the right. This moves the body out of the line of attack and inside the opponent's strike.

Block on two points of contact to disperse the impact of the attack:

Block the opponent's stick by positioning the stick so it is perpendicular to the opponent's stick.

Block the opponent's forearm with the meaty portion of the left forearm.

Note: If closer to the opponent, block his triceps with the back of the left forearm and strike his forearm with the stick.

7. Unarmed Against Hand-Held Weapons

If Marines are engaged against an opponent with a knife, a stick, or some other weapon of opportunity, they must establish and maintain an offensive mindset, not a defensive mindset. Their survival depends on it. They cannot afford to think about getting cut or hurt.

Angles of Attack

Before Marines can learn to block or counter an attack with a hand-held weapon (e.g., knife, stick), they must know from what angle the opponent is attacking. There are six angles from which an opponent will typically attack with a hand-held weapon:

Vertical strike coming in straight down on the Marine.
Forward diagonal strike coming in at a 45-degree angle to the Marine.
Reverse diagonal strike coming in at a 45-degree angle to the Marine.
Forward horizontal strike coming in parallel to the ground.
Reverse horizontal strike coming in parallel to the ground.
Forward thrust coming in a straight line to the Marine.

Blocks

If the opponent has a hand-held weapon (e.g., knife or stick), Marines parry the opponent's hand or arm to block the attack.

Basic Block. To execute the basic block technique, Marines-

Step forward at a 45-degree angle to move out of the line of the attack. Always step in the direction of the strike.

Thrust the forearms forward, hands up, against the opponent's attacking arm. Contact is made on the opponent's arm with the backs of the fore-arms.

Block for a Vertical Strike. To execute the block against a vertical strike, Marines—

Step forward with the left foot at a 45-degree angle to the left to move out of the line of attack.

Thrust the forearms forward, hands up, against the outside of the opponent's attacking arm.

Block for a Forward Strike. To execute the block against a forward diagonal or forward horizontal strike, Marines-

Step forward with the left fool inside the opponent's attacking arm.

Block the attack with both arms bent so the fore-arms make contact with the opponent's biceps and forearm.

Block for a Reverse Strike. To execute the block against a reverse diagonal or reverse horizontal strike, Marines—

Step forward with the right foot to the outside of the opponent's attacking arm.

Block the attack with both arms bent so the fore-arms make contact with the opponent's triceps and forearm.

Block for a Forward Thrust. To execute the block against a forward thrust, Marines—

Bend at the waist, move the hips backward, and jump backward with both feet to move away from the attack. This action is known as "hollowing out."

Hollow out and block the attack with the arms bent and hands together on top of the attacking arm.

Overlap the hands slightly so one thumb is on top of the other hand's index finger. The other thumb should be under the other hand's index finger.

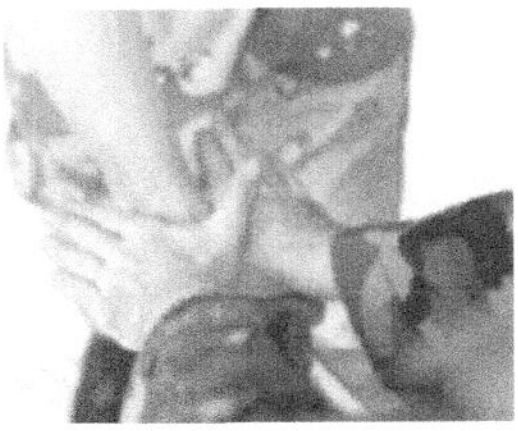

8. Counters to Hand-Held Weapon Attacks

Principles of Counters

A counter is used to control the situation to regain the tactical advantage and end the fight. Regardless of the type of weapon or angle of attack, the following principles apply to countering the attack with a hand-held weapon:

Move out of the line of attack. Movement is executed in a 45-degree angle forward to the left or right.

- Block the attack.

Note: The first two actions are taken simultaneously.

- Take control of the weapon by controlling the hand or arm that is holding the weapon. Never attempt to grab the opponent's weapon.
- Execute the appropriate follow-up to end the fight; e.g., strikes, joint manipulations, throws, or takedowns (see chap. 8). Marines should continue to attack the opponent until the fight ends.

Counter Techniques

There are two techniques that can be used to counter any armed attack: forward armbar counter and reverse armbar counter. These techniques can be used to counter a vertical attack, a forward diagonal strike, or a forward horizontal strike. With minor variations, the same techniques arc used to counter reverse strikes. A third technique, the bent armbar counter, is used to counter a vertical attack.

Forward Armbar Counter. To execute the forward armbar counter to an attack coming from a forward strike, Marines-

Step forward with the left foot inside the opponent' s attacking arm.

Block the attack with both arms bent so the forearms make contact with the opponent's biceps and forearm.

Slide the left arm over the opponent's forearm and wrap the arm tightly around his arm, trapping the opponent's attacking arm between the biceps and torso.

Place the right hand on the opponent's shoulder or upper arm to further control his arm and to effect an armbar.

Execute an armbar and continue to exert steady pressure against the arm.

Reverse Armbar Counter. To execute the reverse armbar counter to an attack coming from a forward strike, Marines-

Step forward with the left foot inside the opponent' s attacking arm.

Block the attack with both arms bent so the forearms make contact with the opponent's biceps and forearm.

Control the opponent's arm with the left hand and pivot to the right so the back is against the opponent's side. Immediately slide the right arm over the opponent's biceps and wrap the arm tightly around his arm, trapping the opponent's attacking arm between the biceps and torso.

Grasp the opponent's wrist with the left hand and twist his thumb away from the body.

Take the opponent to the ground with an armbar takedown if he does not drop the weapon from the wristlock.

Bent Armbar Counter. This counter is particularly effective against a vertical attack. To execute the bent armbar counter, Marines—

Step forward with the left foot inside the opponent's attacking arm.

Block the attack with both arms bent so the forearms make contact with the opponent's biceps and forearm.

Grasp the opponent's forearm with the left hand. At the same time, slide the right arm underneath the opponent's triceps and grasp the opponent's forearm or wrist with the right hand.

Apply downward pressure with the hands against the opponent's forearm to off-balance the opponent.

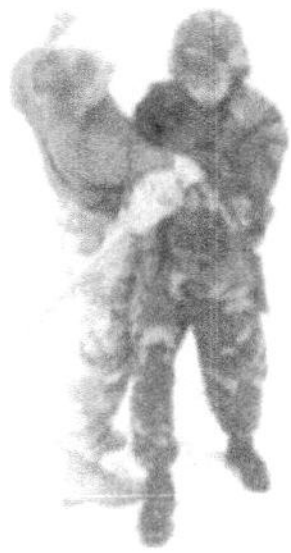

(reverse blank)

CHAPTER 4

STRIKES

This chapter describes all techniques for a right-lmnded person. However, all techniques can be executed from either side.

In drawings, the Marine is depicted in woodland camouflage utilities; the opponent is depicted without camouflage. In photographs, the Marine is depicted in woodland camouflage utilities; the opponent is depicted in desert camouflage utilities.

Strikes are unarmed individual hitting techniques. Strikes use the hands, elbows, knees, feet, and, in some instances, other parts of the body as personal weapons. Marines must know how to execute strikes effectively. They must also know how to counter strikes from an opponent.

1. Principles of Punches

Muscle Relaxation

Muscle relaxation is crucial when executing punches. The natural tendency in a fight is to tense up, which results in rapid fatigue and decreased power generation. Marines who remain relaxed during a close combat situation generate greater speed, which results in greater generation of power. Relaxing the forearms generates speed and improves reaction time. At the point of impact, Marines clench the fist to cause damage to the opponent and avoid injury to the wrist and hand.

Weight Transfer

Weight transfer is necessary to generate power in a punch. Marines accomplish this by-

Rotating their hips and shoulders into the attack.

Moving their body mass straight forward or backward in a straight line.

Dropping their body weight into an opponent. Body mass can be transferred into an attack from high to low or from low to high.

Rapid Retraction

When Marines deliver a punch, rapid retraction of the fist is important. Once the hand has made contact with the target, Marines quickly return to the basic warrior stance. Rapid retraction-

Returns the hand and arm to the protection afforded by the basic warrior stance.

Prevents the opponent from being able to grab the hand or arm.

Permits the hand and arm to be "chambered" or "re-cocked" in preparation for delivering a subsequent punch.

Telegraphing

Telegraphing a strike occurs when body movements inform the opponent of the intention to launch an attack. Staying relaxed helps to reduce telegraphing.

Often, an untrained fighter telegraphs his intention to attack by drawing his hand back in view of his opponent, changing facial expression, tensing neck muscles, or twitching. These movements, however small, immediately indicate an attack is about to be delivered. If the opponent is a trained fighter, he may be able to evade or counter the attack. If the opponent is an untrained fighter, he may still be able to minimize the effect of an attack.

2. Punches

Punches may be thrown during any hand-to-hand confrontation. Most people resort to punching because it is a natural reaction to a threat. The purpose of a punch is to stun the opponent or to set him up for a follow-up finishing technique. However, punches should only be executed to the soft tissue areas of an opponent. A correctly delivered punch maximizes the damage to an opponent while minimizing the risk of injury to Marines.

Basic Fist

Punches are executed using the basic fist. To make the basic fist, the fingers are curled naturally into the palm of the hand and the thumb is placed across the index and middle fingers. Do not clench the fist until movement has begun. This reduces muscular tension in the forearm and increases speed and reaction time. Just before impact, Marines exert muscular tension on the hand and forearm to maximize damage to the opponent and reduce their chances of injury. Contact should be made with the knuckles of the index and middle fingers.

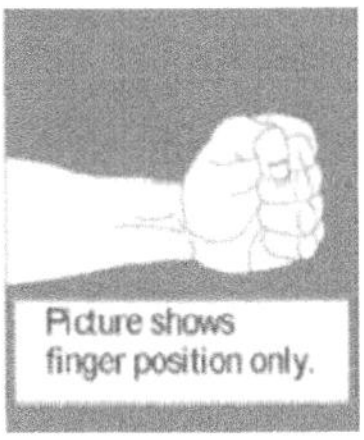

When striking with the basic fist, Marines must keep the hand straight, or in line, with the wrist to avoid injury to the wrist.

Lead Hand Punch

The lead hand punch is a snapping straight punch executed by the forward or lead hand. It is a fast punch designed to keep the opponent away and to set up an attack. A lead hand punch conceals movement and allows Marines to get close to the opponent. Lead hand punches should strike soft tissue areas, if possible. To execute the lead hand punch, Marines-

Snap the lead hand out to nearly full extension, while rotating the palm to the ground.

Contact the opponent with the first two knuckles of the fist.

Retract the hand immediately, resuming the basic warrior stance.

Rear Hand Punch

The rear hand punch is a snapping punch executed by the rear (right) hand. It is a power punch designed to inflict maximum damage on the opponent. Its power comes from pushing off the rear leg and rotating the hips and shoulders. To execute the rear hand punch, Marines —

Rotate the hips and shoulders forcefully toward the opponent and thrust the rear hand straight out, rotating the palm down, to nearly full extension.

Shift body weight to the lead foot while pushing off on the ball of the rear foot.

Contact the opponent with the first two knuckles of the fist.

Retract the hand immediately.

Uppercut

The uppercut is a powerful punch originating below the opponent's line of vision. It is executed in an upward motion traveling up the centerline of the opponent's body. It is delivered in close and usually follows a preparatory strike that leaves the target area unprotected. When delivered to the chin or jaw, the uppercut can render an opponent unconscious, cause extensive damage to the neck, or sever the tongue. To execute the uppercut, Marines—

Bend the arms, rotating the palm inboard. The distance the arms bend depends on how close the opponent is.

Rotate the hips and shoulders forcefully toward the opponent, thrusting the fist straight up toward the opponent's chin or jaw.

Contact the opponent with the first two knuckles of the fist.

Retract the hand immediately.

Hook

The hook is a powerful punch that is executed close in and is usually preceded by a preparatory strike. To execute the hook, Marines—

Thrust the right arm in a hooking motion toward the opponent, keeping the elbow bent while forcefully rotating the right shoulder and hip toward the opponent.

Contact the opponent with the first two knuckles of the fist. Continue rotating the shoulder and hip, following through with the fist to the target.

Retract the hand immediately.

3. Strikes with the Upper Body

Strikes stun the opponent or set him up for a follow-up finishing technique. The hands, forearms, and elbows are individual weapons of the arms that can be used to execute strikes including the hammer fist, chin jab, knife hand, eye gouge, and elbow strikes. These strikes provide a variety of techniques that can be used in any type of close combat situation.

Principles of Execution

There are several principles of execution that ensure a strike's effectiveness.

Generating Power. Weight transfer is necessary to generate maximum power in a punch. Marines accomplish this by-

Rotating the hips and shoulders into the attack.
Moving the body mass straight forward or backward in a straight line.
Dropping body weight into an opponent.
Body mass can be transferred into an attack from high to low or from low to high.

Muscular Tension. The arms are relaxed until the moment of impact. At the point of impact, Marines apply muscular tension in the hand and forearm to maximize damage to the opponent and to avoid injury to the hand. The arms are relaxed until the moment of impact.

Hit and Stick and Follow-Through. A strike should be delivered so that the weapon (e.g., hand, elbow) hits and remains on the impact site (target) and follows through the target. This technique inflicts maximum damage on the opponent.

Strikes with the arms are executed with "heavy hands"; i.e., the strike is executed by driving through with the strike to allow the weight of the hand to go through the target area of the body. Contact with the opponent should be made with the arm slightly bent, and the arm extends as it moves through the target. This technique allows Marines to deliver strikes effectively without executing full force.

Movement. Movement puts Marines in the proper position for launching an attack against an opponent as well as providing protection.

Movement is initiated from the basic warrior stance and ends by resuming the basic warrior stance. Strikes can be performed with either the left or right arm depending upon—

The angle of attack.
The position of the opponent.
The opponent's available, vulnerable target areas.

Target Areas of the Body

For each strike, there are target areas of the body that, when struck, maximize damage to an opponent. Strikes use gross motor skills as opposed to fine motor skills. The target areas of the body are just that—areas. Pinpoint accuracy on a specific nerve is not needed for the strike to be effective.

Hammer Fist

Striking with the hammer fist concentrates power in a small part of the hand, which, when transferred to the target, can have a devastating effect. The striking surface of the hammer fist is the meaty portion of the hand below the little finger. To execute the hammer fist strike, Marines-

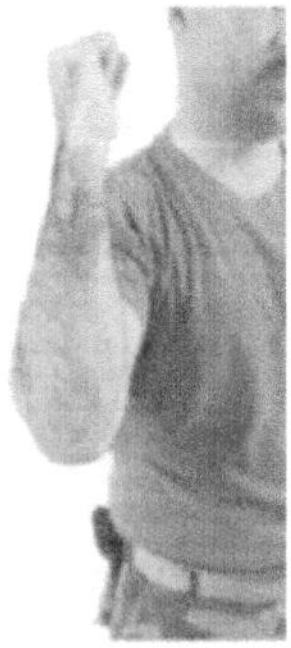

Make a fist and bend the arm at approximately a 45- to 90-degree angle. At the same time, rotate the right hip and right shoulder backward.

Thrust the fist forward onto the opponent while rotating the right hip and shoulder forward.

Rotate the wrist so the hammer fist makes contact on the opponent.

Chin Jab

The chin jab can immediately render an opponent unconscious and cause extensive damage to the neck and spine. The striking surface is the heel of the palm of the hand. To execute the chin jab, Marines—

Bend the right wrist back at a 90-degree angle with the palm facing the opponent and the fingers pointing up.

Keep the right arm bent and close to the body. Extend the hand into a concave position with the fingers slightly spread apart.

Step forward with the left foot toward the opponent, keeping the feet approximately shoulder-width apart and the knees bent. This is done to close with the opponent.

Keep the right arm bent and close to the side. Thrust the palm of the hand directly up under the opponent's chin. At the same time, rotate the right hip forward to drive the body weight into the attack to increase the power of the strike. The attack should travel up the centerline of the opponent's chest to his chin.

Knife Hand

The knife hand is one of the most versatile and devastating strikes. The striking surface is the cutting edge of the hand, which is the meaty portion of the hand below the little finger extending to the top of the wrist. The striking surface is narrow, allowing strikes on the neck between the opponent's body armor and helmet. The knife hand strike is executed from one of three angles: outside, inside, and vertical.

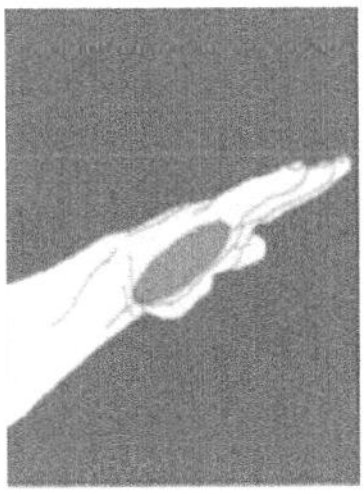

Outside Knife Hand. To execute the outside knife hand strike, Marines—

Execute a knife hand by extending and joining the fingers of the right hand and placing the thumb next to the forefinger (like saluting).

Retract the right hand. At the same time, rotate the right hip and right shoulder backward.

Thrust the knife hand forward (horizontally) onto the opponent while rotating the right hip and shoulder forward.

Inside Knife Hand. To execute the inside knife hand strike, Marines—

Execute a knife hand.

Bring the right hand over the left shoulder. At the same time, rotate the right shoulder forward and the left hip forward.

Thrust the knife hand forward (horizontally) onto the opponent while rotating the right hip and shoulder forward and the left shoulder backward.

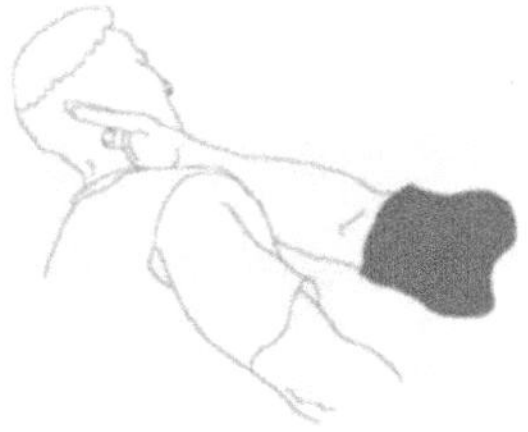

Vertical Knife Hand. When thrown vertically, the knife hand strike comes straight down in a straight line.

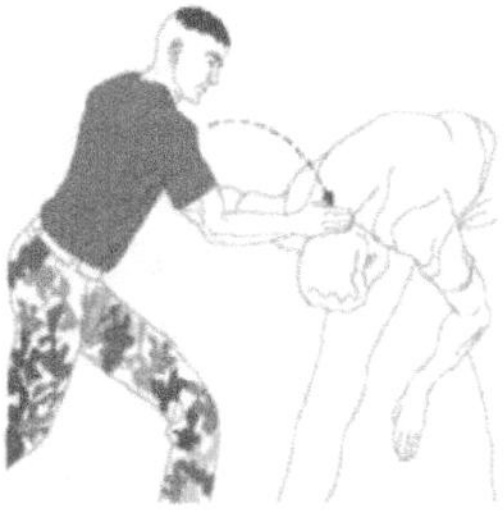

Eye Gouge

The eye gouge is used to attack an opponent's eyes, blinding him so follow-up strikes can be executed. The striking surface is the tips of the fingers or thumb. To execute the eye gouge, Marines—

Extend the right hand with the fingers slightly spread apart to allow entry into the eye sockets.

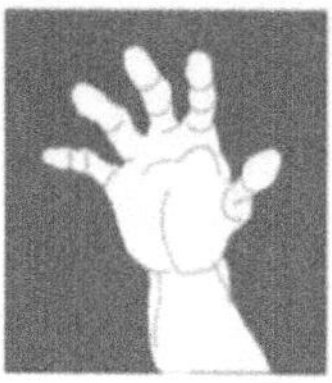

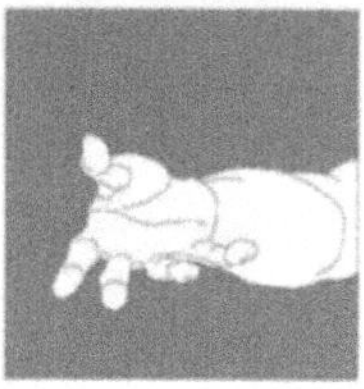

Place the palm of the hand either toward the ground or toward the sky and thrust the right hand forward into the opponent's eyes.

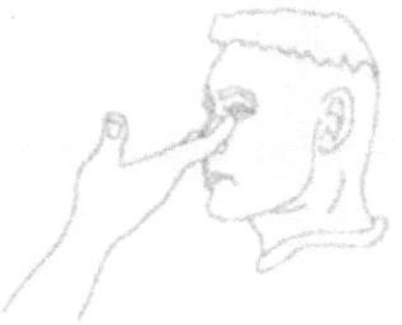

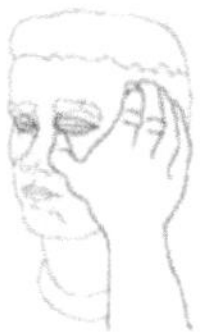

Thrust the hand forward at the opponent's nose level so the fingers or thumb slide naturally into the grooves of the opponent's eye sockets.

Elbow Strikes

The elbow is a powerful weapon that can be used in several different ways to attack virtually any part of an opponent's body. Elbow strikes can be performed either vertically (upward or downward) or horizontally (forward or reverse). The striking surface is 2 inches above or below the point of the elbow, depending upon the angle of attack, the opponent's attack angle, and the position of the opponent.

Above Elbow

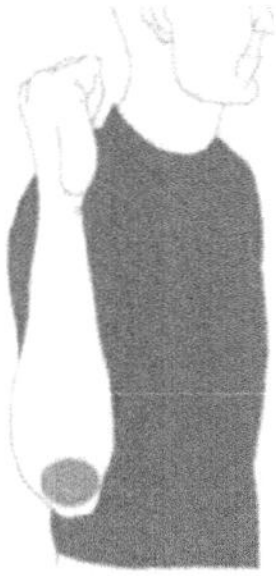
Below Elbow

Vertical Elbow Strike (Up). To execute an upward vertical elbow strike, Marines-

Bend the right elbow, keeping the fist close to the body. The fist is at shoulder level and the elbow is next to the torso.

Thrust the elbow vertically upward toward the opponent while rotating the right shoulder and hip forward to generate additional power.

Contact the opponent with the right forearm 2 inches above the point of the elbow.

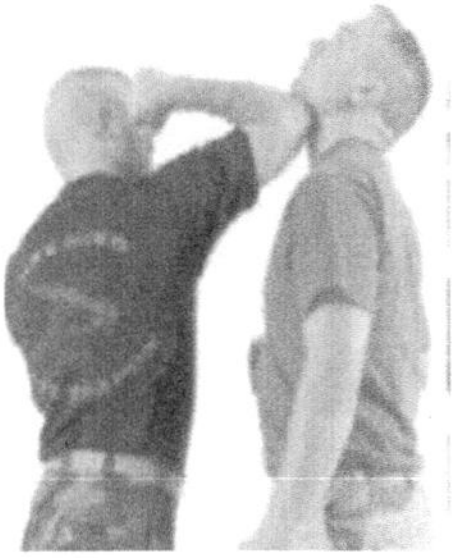

Vertical Elbow Strike (Down). To execute a downward vertical elbow strike, Marines—

Bend the right elbow, keeping the fist close to the body. The fist is on the shoulder and the elbow is raised well above the shoulder.

Thrust the elbow vertically downward toward the opponent while dropping body weight into the attack to generate additional power.

Contact the opponent with the right triceps 2 inches above the point of the elbow.

Horizontal Elbow Strike (Forward). To execute the forward horizontal elbow strike, Marines-

Tuck the right fist near the chest with the palm heel facing the ground.

Thrust the right elbow horizontally forward toward the opponent. The forearm is parallel to the ground.

Rotate the right shoulder and hip forward to generate additional power.

Contact the opponent with the right forearm 2 inches below the point of the elbow.

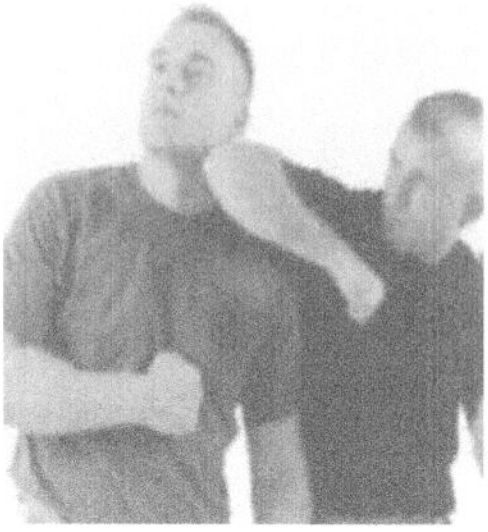

Horizontal Elbow Strike (Rear). To execute the rear horizontal elbow strike, Marines-

Tuck the right fist near the left shoulder with the palm heel facing the ground. At the same time, rotate the right shoulder forward and the left hip forward.

Thrust the right elbow horizontally rearward toward the opponent. The forearm is parallel to the ground and the hand moves toward the direction of the attack.

Rotate the right hip back and the right shoulder backward to generate additional power.

Contact the opponent with the right triceps 2 inches above the point of the elbow.

4. Strikes with the Lower Body

The legs are the body's most powerful weapons because they use the largest muscle groups to generate a strike. Legs are also less prone to injury. The feet are the preferred choice for striking because they are protected by boots. Marines use their feet, heels, and knees to execute kicks, knee strikes, and stomps.

Kicks

The purpose of a kick is to stop an opponent's attack or to create an opening in his defense in order to launch an attack. Kicks can be performed with the left (lead) leg or the right (rear) leg. Kicks with the rear leg have greater power because the hips are rotated into the attack. However, the rear leg is further away from the opponent so a strike with the rear leg will not contact the opponent as quickly as a strike with the lead leg.

Front Kick. The front kick is executed when the opponent is in front of the Marine. The front kick, delivered with either the rear or lead leg, is effective for striking below the waist. Attempting to kick higher results in diminished balance and provides the enemy with a greater opportunity to grab the leg or foot. The striking surfaces arc the toe of the boot or the bootlaces. To execute the front kick, Marines-

Raise the left knee waist high, pivot the hips into the attack, and thrust the left foot forward toward the opponent.

Contact the opponent with the toe of the left boot or bootlaces.

Return to the basic warrior stance.

Side Kick. The side kick, delivered with the lead leg, is effective for striking the knees. The side kick is executed when the opponent is to the side of the Marine. The striking surface is the outside cutting edge of the boot near the heel. To execute the side kick, Marines—

Raise the right knee waist high and rotate the right hip forward.

Thrust the right foot to the right side toward the opponent, turning the foot at a 90-degree angle to maximize the striking surface on the opponent.

Contact the opponent with the cutting edge of the right boot.

Return to the basic warrior stance.

Knee Strikes

Knee strikes are excellent weapons during the close range of close combat fighting. The knee strike is generally delivered in close.

Vertical Knee Strike. The striking surface is the thigh, slightly above the knee. To execute the vertical knee strike, Marines-

Raise the right knee and drive it up forcefully into the opponent. Power is generated by thrusting the leg upward.

Contact the opponent 2 inches above the right knee.

Horizontal Knee Strike. The horizontal knee strike is executed with the leg generally parallel with the ground while rotating the hips to generate power. It is often delivered to the peroneal nerve. The striking surface is the front of the leg, slightly above or below the knee. To execute the horizontal knee strike, Marines-

Raise the right knee, rotate the right hip forward while pivoting on the left foot, and drive the knee horizontally into the opponent.

Contact the opponent 2 inches above the right knee.

Stomps

Stomps are delivered with the feet, usually when the opponent is down. Remember, when the opponent is down, Marines take whatever target is available.

Vertical Stomp. The vertical stomp allows Marines to remain upright and balanced, to rapidly deliver multiple blows with either foot, and to quickly and accurately attack the target. It is the preferred stomp. The striking surface is the flat bottom of the boot or the cutting edge of the heel. To execute the vertical stomp, Marines—

Raise the right knee above the waist with the right leg bent at approximately a 90-degree angle.

Drive the flat bottom of the right boot or the cutting edge of the right heel down onto the opponent forcefully. At the same time, bend the left knee slightly to drop the body weight into the strike.

Ax Stomp. The striking surface of the ax stomp is the cutting edge of the heel. To execute the ax stomp, Marines-

Raise the right heel above the waist, keeping the right leg straight.

Drive the cutting edge of the right heel down onto the opponent forcefully. At the same time, bend the left knee slightly to drop the body weight into the strike.

5. Counters to Strikes

In a close combat situation, an opponent will attempt to strike Marines with punches and kicks. When an opponent uses a strike, Marines must first avoid the strike. This is accomplished by moving quickly and blocking. Next, Marines must get into an offensive position. This allows Marines to use offensive strikes to attack the opponent. Regardless of the strike, the counter to a strike requires Marines to move, block, and strike.

Move

The first step in countering a strike is to move out of the way of the strike's impact. Movement removes Marines from the opponent's intended strike point and positions Marines to attack. Movement is executed at approximately a 45-degree angle to the front or rear. Movement is always initiated from the basic warrior stance. Return to the basic warrior stance with the toe of the lead foot pointing toward the opponent once the movement is complete.

Block

Different blocks are executed based on the strike. These will be covered by individual counters.

Strike

Any of the upper body or lower body strikes can be executed as a follow-on attack or part of the counter to an opponent's strike. The follow-on strike is determined by the angle to the opponent, the position of the opponent, and the opponent's available vulnerable target areas.

Counters to Punches

Counter to a Lead Hand Punch. This counter is used when the opponent throws a lead hand punch. To execute the counter to the lead hand punch, Marines-

Step forward and to the left at approximately a 45-degree angle, moving in to the outside of the opponent's attacking arm.

Raise the left arm and block or deflect the opponent's lead hand with the palm of the hand or the meaty portion of the forearm.

Hit and stick by leaving the left arm against the opponent's right arm while stepping forward and to the left at approximately a 45-degree angle to close with the opponent.

Execute a finishing technique, such as a strike or a kick, to the opponent's exposed target areas.

Counter to a Rear Hand Punch. This counter is used when the opponent throws a rear hand punch. To execute the counter to the rear hand punch, Marines—

Step forward and to the left at approximately a 45-degree angle, moving in to the outside of the opponent's attacking arm.

Raise the left arm and block or deflect the opponent' s rear hand with the palm of the hand or the meaty portion of the forearm.

Hit and stick by leaving the left arm against the opponent's right arm while stepping forward and to the right at approximately a 45-degree angle to close with the opponent.

Execute a finishing technique, such as a strike or a kick, to the opponent's exposed target areas.

Counters to Kicks

Counter to a Front Kick (Left or Lead Leg). This counter is used when the opponent executes a front kick with his left leg. To execute the counter to a front kick, Marines-

Step forward and to the right at approximately a 45-degree angle, moving in to the outside of the opponent's striking leg.

Lower the left arm and block or deflect the opponent' s leg with the palm of the hand or the meaty portion of the forearm.

Hit and stick by leaving the left arm against the opponent's leg while stepping forward and to the left at approximately a 45-degree angle to close with the opponent.

Execute a finishing technique, such as a strike or a kick, to the opponent's exposed target areas.

Counter to a Front Kick (Right or Rear Leg). This counter is used when the opponent executes a front kick with his right leg. To execute the counter to a front kick, Marines—

Step forward and to the left at approximately a 45-degree angle, moving in to the outside of the opponent's striking leg.

Lower the left arm and block or deflect the opponent' s leg with the palm of the hand or the meaty portion of the forearm.

Hit and stick by leaving the left arm against the opponent's leg while stepping forward and to the right at approximately a 45-degree angle to close with the opponent.

CHAPTER 5

THROWS

This chapter describes all techniques for a right-handed person. However, all techniques can be executed from either side.

The Marine is depicted in camouflage utilities. The opponent is depicted without camouflage.

Marines use throwing techniques to maintain the tactical advantage and to throw the opponent to the ground during close combat. Throws apply the principles of balance, leverage, timing, and body position to upset an opponent's balance and to gain control by forcing the opponent to the ground. Throwing techniques are effective because they are size- and gender-neutral, and they rely on the momentum and power generated by the opponent rather than the strength or size of the Marine. Marines also execute a throw as a devastating attack against an opponent, possibly causing unconsciousness or broken limbs. When Marines execute throws, they must maintain balance and, simultaneously, prevent the opponent from countering a throw or escaping after being forced to the ground.

1. Turning Throw

Marines use a turning throw to take the opponent to the ground while they remain standing. A turning throw can also be executed from a stationary position. It is particularly effective if Marines and the opponent are wearing gear. To execute the turning throw, Marines-

Grasp the opponent's right wrist with the left hand.

Step forward with the right foot, place it against the outside of the opponent's right foot, and pivot so the back of the heel is next to the middle of the opponent's foot.

Hook the opponent's right arm with the right arm and pinch his arm between the biceps and forearm, touching the opponent with the body.

Pull the opponent's wrist downward, keeping it close to the body.

Pivot to the left on the ball of the foot and continue pulling downward on the opponent's wrist while rotating the wrist outward to off-balance him.

2. Hip Throw

Marines use a hip throw to take the opponent to the ground while they remain standing. A hip throw is particularly effective if the opponent is moving forward or pushing. Marines use the opponent' s forward momentum to execute the hip throw. To execute the hip throw, Marines-

Grasp the opponent's right wrist with the left hand.

Step forward with the right foot and place it against the outside of the opponent's right foot.

The back of the heel should be next to the opponent' s foot.

Step back with the left foot and rotate on the ball of the foot. The back of the heel is next to the opponent' s toe. The knees are bent.

Rotate at the waist and hook the right hand around the back of the opponent's body (anywhere from his waist to his head). The side and hip should be against the opponent.

Rotate the hip against the opponent. The hips must be lower than the opponent's.

Use the right hand to pull the opponent up on the hip to maximize contact.

Pull the opponent's arm across the body and, at the same time, lift the opponent off the ground slightly by bending at the waist, straightening the legs, and rotating the body to the left.

3. Leg Sweep

Marines use a leg sweep to take the opponent to the ground while they remain standing. A leg sweep is particularly effective if the opponent is already off-balanced and moving backward or pulling on the Marine. To execute the leg sweep, Marines—

Grab the opponent's right wrist with the left hand and grab the opponent' s left shoulder with the right hand.

Note: Grab the opponent's clothing or gear if his wrist and shoulder cannot be grabbed.

Step forward with the left foot and place it on the outside of the opponent's right foot.

Pull the opponent's wrist downward, close to the body, and push his shoulder backward to off-balance him.

Raise the right knee no higher than the waist.

Kick the foot past the opponent's right leg.

Use the heel of the boot to make contact with the opponent's calf (anywhere from the top of the calf down to the Achilles tendon or on the inside of the calf).

Sweep the opponent to the ground.

CHAPTER 6

CHOKES AND HOLDS

This chapter describes all techniques for a right-handed person. However, all techniques can be executed from either side.

The Marine is depicted in camouflage utilities. The opponent is depicted without camouflage.

WARNING

During training, never execute a choke at full force or full speed and never hold a choke for more than 5 seconds.

When Marines correctly perform a choke, they render an opponent unconscious in as little as 8 to 13 seconds. Chokes are easily performed regardless of size or gender. Marines must know how to apply chokes and how to counter a choke or a hold executed by an opponent.

1. Types of Chokes

There are two types of chokes: an air choke and a blood choke. An air choke closes off the airway to the lungs, thereby preventing oxygen from reaching the heart. A blood choke cuts off the blood flow to the brain. Both types can result in unconsciousness and eventual death for an opponent.

Air Choke

An air choke is performed on the opponent's windpipe (or trachea), cutting off the air to the lungs and heart. If Marines execute the air choke properly, the opponent loses consciousness within 2 to 3 minutes. Due to the length of time it takes to immobilize the opponent, air chokes are not recommended.

Blood Choke

A blood choke is performed on the opponent's carotid artery, which carries oxygen-enriched blood from the heart to the brain. The carotid artery is located on both sides of the neck. If Marines execute a blood choke properly, the opponent will lose consciousness within 8 to 13 seconds. The blood choke is the preferred choke because its intended effect (i.e., the opponent losing consciousness) can be executed quickly, ending the fight.

2. Chokes

Front Choke

Marines execute a front choke when they are facing the opponent. Marines use the opponent's lapels or collar to execute a front choke. To execute the front choke, Marines-

Grab the opponent's right lapel with the right hand, making certain that the knuckles or the back of the hand are against the opponent's carotid artery on the right side of his neck.

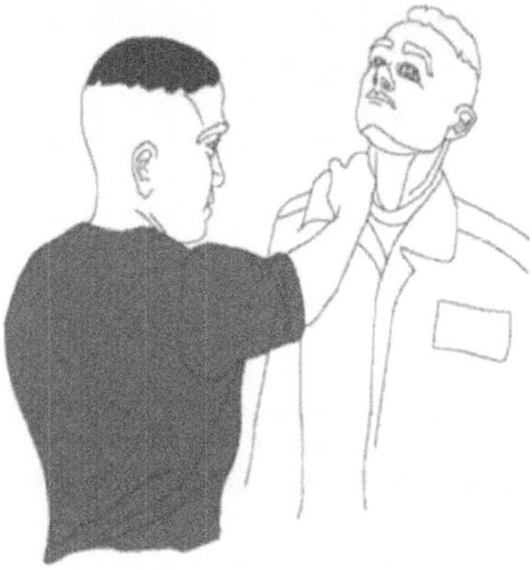

Keep the right hand against the opponent's neck, reach under the opponent's right arm with the left hand, grab the opponent's left lapel, and form an X with the wrists.

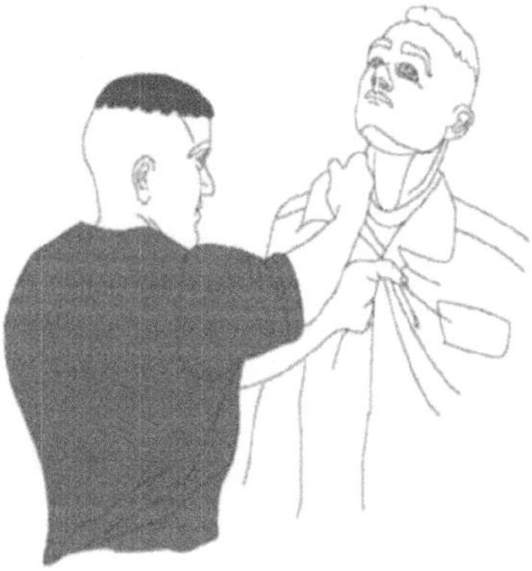

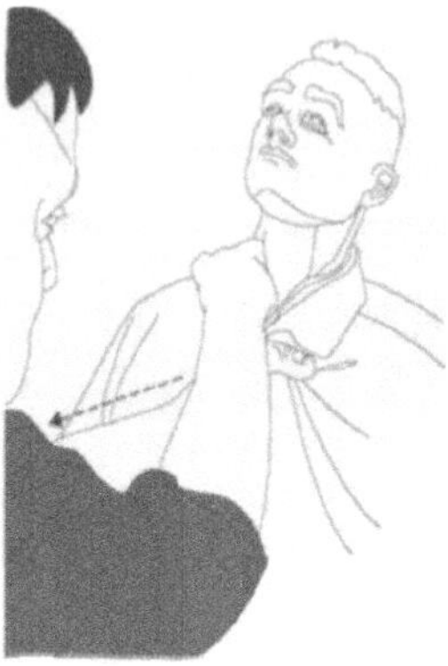

Press the right hand against the opponent's carotid artery.

Pull his lapel to the left with the left hand.

Side Choke

Marines execute a side choke when they are facing the opponent. The side choke is particularly effective when deflecting a punch thrown by an opponent. To execute the side choke, Marines—

Use the left hand to parry the opponent's arm inboard (to the inside of the opponent's reach).

Bring the right arm underneath the opponent's arm and up around the front of his neck.

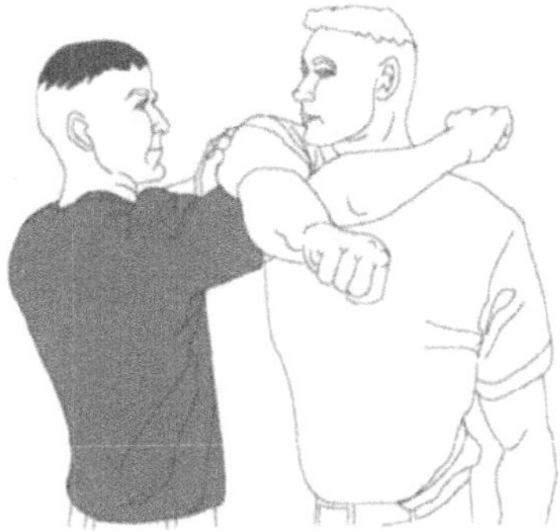

Extend the fingers, place the back of the forearm against the opponent's neck just below his ear, and press the carotid artery on his neck.

Note: If Marines are unable to place their fingers against the opponent's neck, they may use the back of the thumb or wrist.

Reach, with the left hand, around the back of the opponent's neck and clasp the hands together.

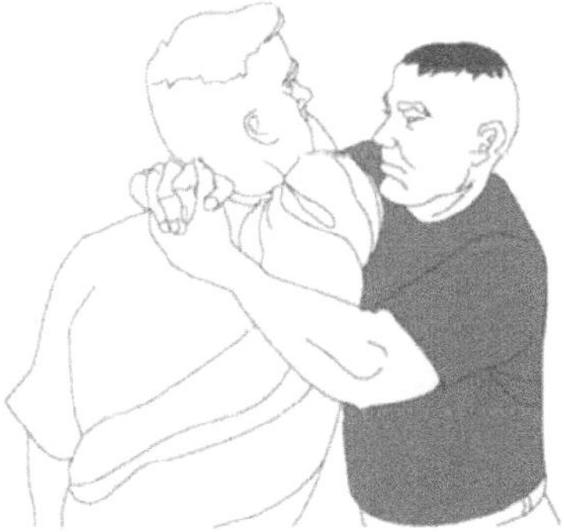

Pull the opponent toward the chest by pulling the clasped hands toward the chest.

Exert pressure on the side of the opponent's neck with the forearm.

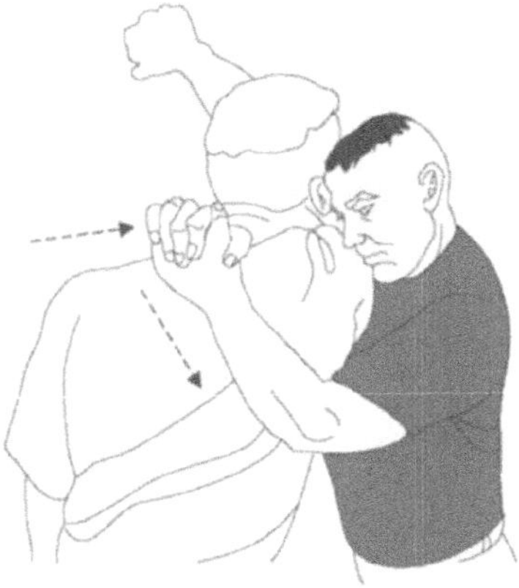

Rear Choke

Marines execute a rear choke when they are behind the opponent, the opponent is on the ground, or when they are taking the opponent to the ground. To execute the rear choke, Marines—

Reach, with the right arm, over the opponent's right shoulder and hook the bend of the arm around his neck.

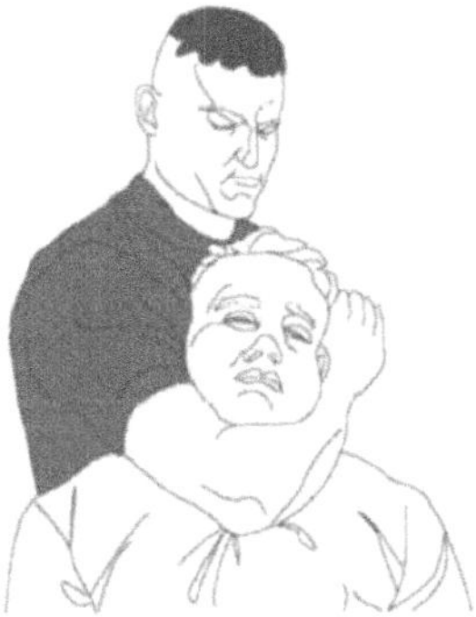

Clasp both hands together.

Exert pressure with the biceps and forearms on both sides of the opponent's neck on his carotid arteries.

Maintain pressure with the biceps and forearms on both sides of the neck and draw the opponent closer by drawing the right arm in.

Figure-Four Choke

A variation of the rear choke is the figure-four choke. The figure-four choke allows Marines to gain more leverage than the rear choke. If the rear choke cannot be secured, the figure-four variation is employed to increase the pressure of the choke on the opponent. To execute the figure-four variation of the rear choke, Marines—

Apply a rear choke. The Marine's body should be against the opponent's body.

Grasp the left biceps with the right hand and place the left hand against the back of the opponent's head.

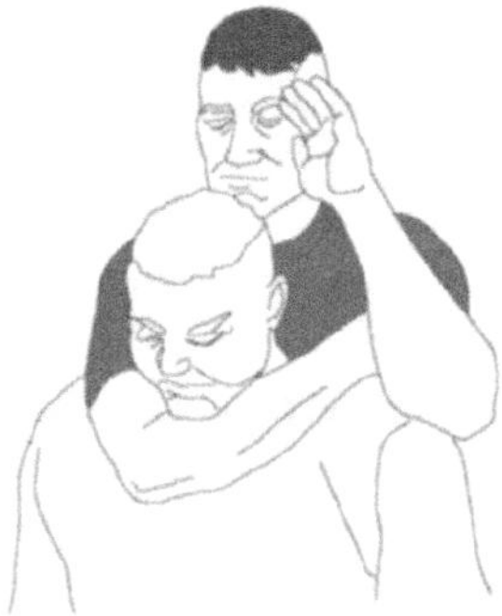

Push the opponent's head forward and down with the left hand.

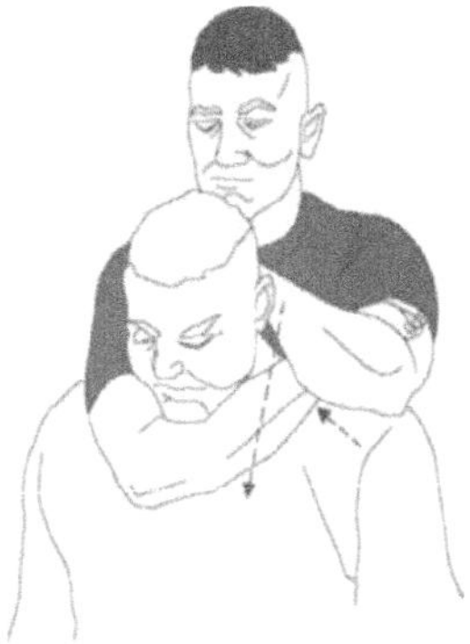

Draw the right arm in, maintaining pressure with the bicep and forearm on both sides of the opponent's neck.

3. Counters to Chokes and Holds

During a close combat situation, an opponent may apply a choke or hold on a Marine. If the opponent correctly applies a choke, a Marine quickly loses consciousness. If a choke is not executed properly, it often results in a hold, typically a bear hug or a headlock. A hold allows the opponent to control a Marine and removes the Marine's ability to attack. It is important for Marines to extract themselves from chokes and holds, regain the tactical advantage, and counter with strikes.

Although a choke causes unconsciousness in 8 to 13 seconds for a blood choke and 2 to 3 minutes for an air choke, the first movement in any counter to a choke is to clear the airway. Marines use softening techniques to loosen an opponent's grip and to clear their airway. Softening techniques are particularly effective if Marines lack the physical strength of their opponent. These techniques include groin strikes, eye gouges, foot stomps, etc. Softening techniques are not offensive; rather, they are used to loosen an opponent's hold.

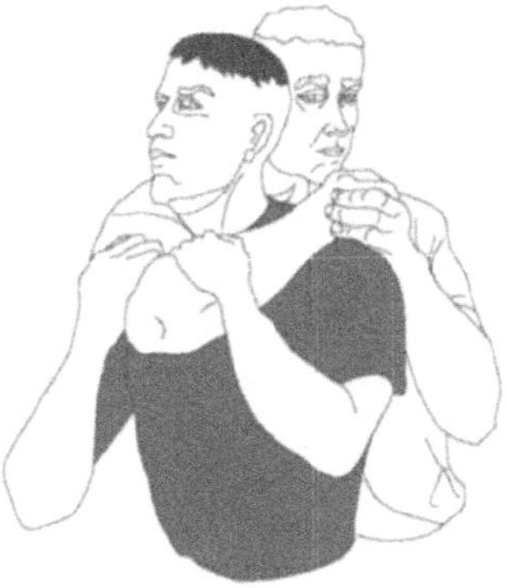

The second movement is to tuck the chin. Once the airway is clear, Marines tuck their chins to prevent the opponent from reapplying the choke.

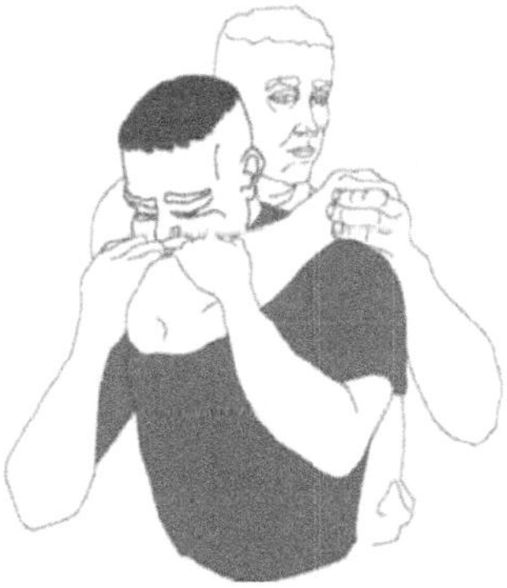

Counter to a Front Choke

Marines use a counter to a front choke when the opponent approaches from the front and uses both hands to choke a Marine around the throat. To execute the counter to the front choke, Marines—

Grasp the opponent's right forearm (where the elbow bends) with the left hand and apply downward pressure on the opponent's radial nerve with the fingers.

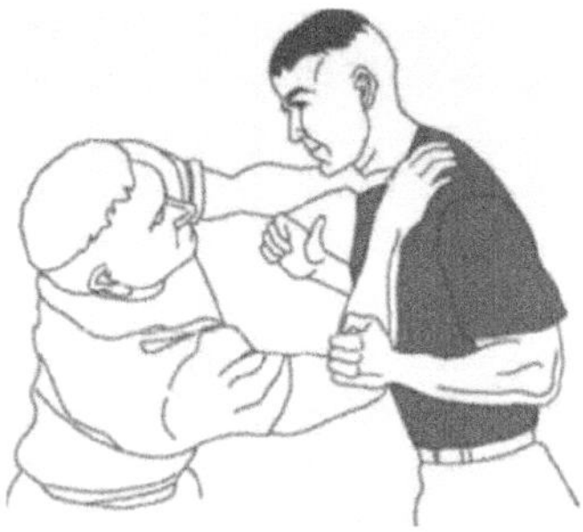

Execute a chin jab to the opponent's chin with the right hand. To generate power into the strike, bring the left foot to the outside of the opponent's right foot and rotate the hips into the strike.

Counter to a Rear Choke

Marines execute a counter to a rear choke when the opponent approaches from the rear and puts his right arm around a Marine's throat. To execute the counter to the rear choke, Marines-

Grasp the opponent's forearm (at the radial nerve) and bicep with both hands and pull down just enough to clear the airway. Once the airway is clear, tuck the chin to protect the airway and to prevent the opponent from reapplying the choke.

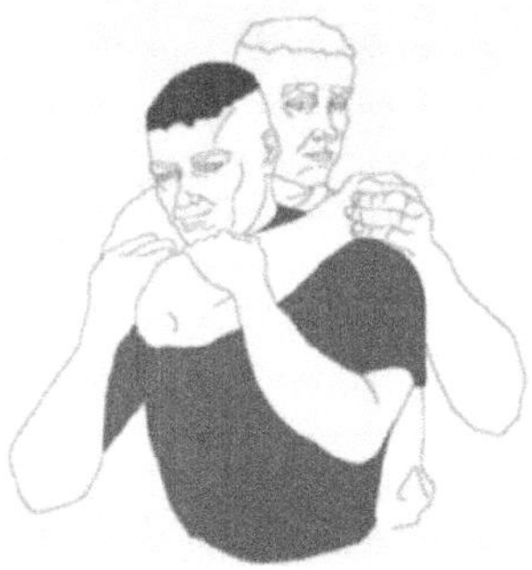

Step behind the opponent's right leg with the left foot, keeping both legs bent (almost in a squatting position).

Strike and drive the left elbow into the opponent's torso while rotating the hips and pivoting to the left, throwing the opponent back and to the ground.

Counter to a Front Headlock

Marines use a counter to a front headlock when the opponent approaches from the front and puts his right arm around the Marine's neck, bends the Marine forward, and locks the Marine's head against his hip. To execute the counter to a front headlock, Marines—

Grasp the opponent's wrist and forearm with both hands and pull down to clear the airway. Maintain control of the opponent's wrist throughout the move. Once the airway is clear, tuck the chin to protect the airway and to prevent the opponent from reapplying the choke.

Move the right hand and arm across the opponent' s torso.

Step forward and to the left with the left foot at a 45-degree angle.

Execute a sweep with the right foot against the opponent's right leg. At the same time, push against the opponent's chest with the right arm and shoulder to generate power in the sweep.

Counter to a Rear Headlock

Marines use a counter to a rear headlock when the opponent approaches from the rear and puts his right arm around the Marine's neck, bends the Marine forward, and locks the Marine's head against his hip. To execute the counter to the rear headlock, Marines-

Grasp the opponent's wrist and forearm with the right hand and pull down to clear the airway. Once the airway is clear, tuck the chin to protect the airway and to prevent the opponent from re-applying the choke.

Reach over the opponent's right shoulder with the left arm.

Grab any part of the opponent's face (chin, nose, eyes) and pull back while rising to a standing position.

Execute, with the right hand, a hammer fist strike to the opponent's exposed throat.

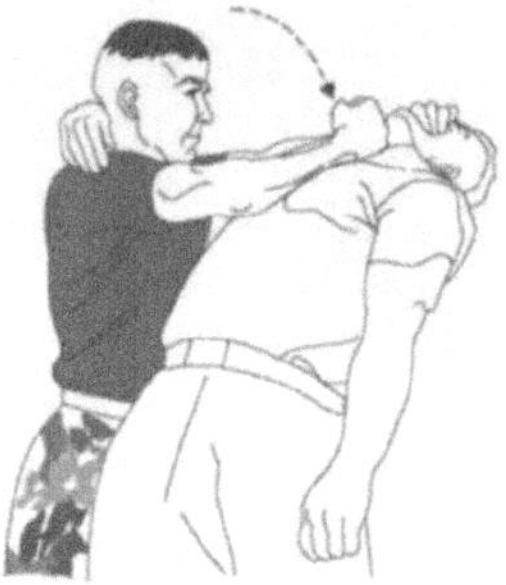

Counter to a Front Bear Hug

Marines execute a counter to a front bear hug when the opponent approaches from the front and puts both of his arms around the Marine's body, trapping the Marine's arms to the sides. To execute the counter to a front bear hug, Marines —

Step forward and to the left with the left foot at a 45-degree angle to the outside of the opponent's right leg, keeping the left leg bent.

Grasp the opponent's torso or arms to gain balance and to assist in throwing the opponent. It may be helpful to hook the opponent's right arm with the left arm.

Drive the right arm and shoulder forward and, at the same time, bring the right leg forward and sweep the opponent's right leg, bringing him to the ground.

Counter to a Rear Bear Hug

Marines use a counter to a rear bear hug when the opponent approaches from the rear and puts both of his arms around the Marine's body, trapping the Marine's arms to the sides. To execute the counter to a rear bear hug, Marines—

Step behind the opponent's right leg with the left foot, keeping both legs bent (almost in a squatting position). The left side of the body should be against the opponent's.

Pivot the hip, turning the body to the left and throwing the opponent back over the bent leg.

(reverse blank)

CHAPTER 7

GROUND FIGHTING

This chapter describes all techniques for a right-handed person. However, all techniques can be executed from either side.

In drawings, the Marine is depicted in woodland camouflage utilities; the opponent is depicted without camouflage. In photographs, the Marine is depicted in woodland camouflage utilities; the opponent is depicted in desert camouflage utilities.

Marines should avoid being on the ground during a close combat situation because the battlefield may be covered with debris and there is an increased risk of injury. However, many close combat situations involve fighting on the ground. The priority in a ground fight is for Marines to get back on their feet as quickly as possible. In any ground fighting scenario, Marines will usually end up in one of four positions with the opponent. The offensive positions, in which Marines have a tactical advantage, are the guard and mount. The defensive positions, which are used as counters when the opponent has the tactical advantage, are the counter to the guard and the counter to the mount. Marines can also employ chokes during ground fighting to quickly end a fight.

1. Offensive Ground Fighting

Guard Position

Marines execute a guard position when the opponent is on top and the legs are wrapped around the opponent's legs. If the opponent is on top, he may try to choke the Marine, but the Marine still has the tactical advantage because the Marine is in a position to control the situation. To execute the guard position technique, Marines—

Trap the opponent's hands on the chest by crossing the hands on the chest so the forearms are resting on the opponent's forearms. Apply pressure with the elbows or forearms.

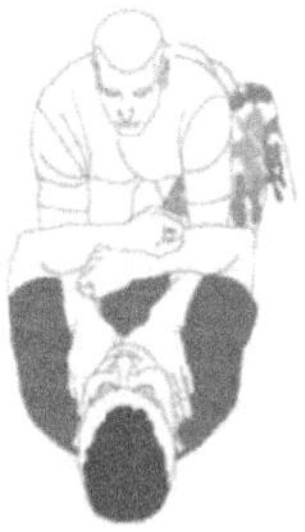

Strike the outside of the opponent's thigh with the cutting edge of the right heel. This causes the opponent to jerk to that side.

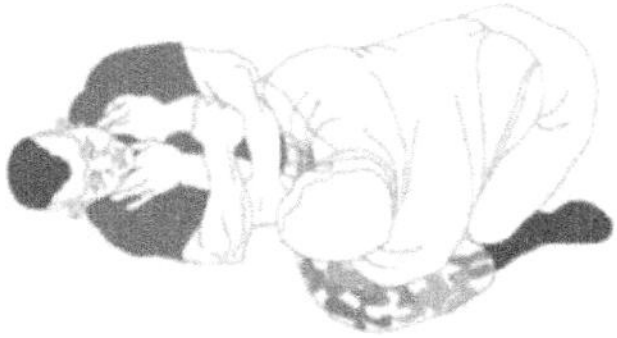

Move the head quickly to the left and swivel the hips to the right. At the same time, bring up both of the legs. Both legs are on the right side of the opponent's body.

Bring the right leg down, hooking the opponent's neck and head. Exert downward pressure to roll him over on his back. Grasp and maintain control of the opponent's left arm. Upon completion of the move, the Marine is sitting up with legs bent over the opponent while maintaining control of his arm.

Keep legs and knees bent. Maintain pressure against the opponent's neck with the back of the right foot and against his side with the left foot underneath his armpit. Squeeze the knees together, locking the opponent's arm.

Pull the opponent's arm straight up and fall back sharply, pulling his arm to the side in the direction of his little finger.

Return to standing.

Mount Position

Marines execute the mount if the opponent is lying on his back on the ground and the Marine is on top with legs wrapped around the opponent's body. This position is an offensive position because Marines are in a better position to control the opponent and to execute ground fighting techniques. To execute the mount position technique, Marines—

Grab the opponent's wrists or forearms with the hands. Hold them tightly against the torso.

Maintain control of the opponent's arms with the left arm and apply pressure to the opponent's brachial plexus (tie in) with the right hand.

Bring the left foot close to the opponent's right armpit and swing the right leg across the opponent' s head.

Keep the legs and knees bent. Maintain pressure against the opponent's neck with the back of the right foot and against his side with the left foot which is underneath his armpit. Squeeze the knees together, locking the opponent's arm.

Pull the opponent's arm straight up and fall back sharply, pulling his arm to the side in the direction of his little finger.

Return to standing.

2. Defensive Ground Fighting

Counter to the Guard

Marines use this technique if the opponent is lying on his back on the ground and the Marine is kneeling on the ground between the opponent's legs. To execute the counter to the guard position, Marines—

Strike the opponent's femoral nerve, located on the inside of the thigh, with the elbows. This forces the opponent to separate his legs.

Strike the opponent's groin with the right fist.

Hook the left arm underneath the opponent's right knee from the inside. Throw the opponent's leg over the head.

Duck the head quickly and move to the left.

Return to standing.

Counter to the Mount

Marines use this technique when they are lying on their back on the ground, the opponent is mounted on top of the Marine, and the opponent's legs are wrapped around the Marine. The opponent has the tactical advantage. To execute the counter to the mount position, Marines—

Grab the opponent's gear or clothing on his upper torso and pull him down close.

Use the right arm to hook the opponent's left arm, from the inside around the outside, above his elbow. With the right foot, hook the opponent's left leg or ankle.

Draw the elbow in to bend the opponent's elbow, bringing him down close. The arm must be hooked above the opponent's elbow in order to bend it.

Strike the opponent's side with the left hand.

Push the opponent over and roll him off to the right side.

Maintain control of the opponent's hooked arm and move to a kneeling position.

Move to a standing position while maintaining pressure on the opponent's arm and using the knee to apply pressure against the opponent's elbow.

3. Ground Fighting Chokes

WARNING

During training, never execute a choke at full force or full speed. Never hold a choke for more than 5 seconds.

The priority in ground fighting is for Marines to get back on their feet as quickly as possible. Sometimes, Marines can quickly end a ground fight by executing a choke on the opponent. When performed correctly, a choke can render an opponent unconscious in as little as 8 to 13 seconds. Chokes are easily performed regardless of size or gender. The chokes performed during ground fighting are the same as those performed while standing (see chap. 6).

Ground Fighting Front Choke

The ground fighting front choke is a blood choke performed most effectively from the mount position. The front choke employs the opponent's lapels or collar to execute the choke. To execute the ground fighting front choke, Marines—

Grab the opponent's right lapel with the right hand, making certain that the knuckles or the back of the hand are against the carotid artery on the right side of the opponent's neck.

Keep the right hand pressed against the opponent's carotid artery, reach under the right arm with the left hand, and grab the opponent's left lapel, forming an X with the wrists.

Keep the right hand pressed against the opponent's carotid artery and pull the opponent's left lapel to the left with the left hand.

Ground Fighting Side Choke

The ground fighting side choke is a blood choke performed from the mount position. The ground fighting side choke is particularly effective when the opponent raises his arms and places them on the Marine's chest or throat. To execute the ground fighting side choke, Marines-

Use the left hand to parry the opponent's right arm inboard (to the inside of the opponent's reach). Bring the right arm underneath the opponent's arm and up around the front of his neck.

Keep the fingers extended, place the back of the forearm against the opponent's neck just below his ear and press the carotid artery on his neck. Depending on the Marine's or the opponent's body configuration, the back of the thumb or wrist may press against the opponent's neck.

Use the left hand to reach around the back of the opponent's neck and clasp the hands together.

Pull the opponent toward the chest. Use the forearm to exert pressure on the side of his neck. This is done by pulling the clasped hands toward the chest.

Ground Fighting Rear Choke

The ground fighting rear choke is a blood choke performed when Marines are behind the opponent. To execute the ground fighting rear choke, Marines—

Wrap the legs around the opponent with the left foot against the inside of his left thigh and the right foot against the inside of his right thigh.

Use the right arm to reach over the opponent's right shoulder and hook the bend of the arm around his neck.

Use the left hand to clasp both hands together.

Exert pressure with the biceps and forearms on both sides of the opponent's neck. This applies pressure to the opponent's carotid arteries. While maintaining pressure with the biceps and forearms, draw the opponent closer by drawing the right arm in. To increase the effectiveness of the choke, lean backward by arching the back and pulling the opponent back. At the same time, push the feet against the opponent's thighs.

Figure-Four Choke

A variation of the ground fighting rear choke is the figure-four choke. The figure-four choke allows Marines to gain more leverage on the rear choke. If the rear choke cannot be secured, the figure-four variation may be applied to increase the pressure of the choke on the opponent. To execute the figure-four variation of the rear choke, Marines—

Apply the ground fighting rear choke. The chest should be against the opponent's back.

Grasp the left biceps with the right hand and place the left hand against the back of the opponent's head.

Use the left hand to push the opponent's head forward and down.

Draw the right arm in while maintaining pressure with the biceps and forearm on both sides of the opponent's neck.

Increase the effectiveness of the choke by arching backward and pushing the feet against the opponent's thighs. Continue to pull the opponent back with the right arm while exerting pressure forward with the left hand against the opponent's head.

CHAPTER 8

NONLETHAL TECHNIQUES

This chapter describes all techniques for a riglit-handed person. However, all techniques can be executed from either side.

In drawings, the Marine is depicted in woodland camouflage utilities; the opponent is depicted without camouflage. In photographs, the Marine is depicted in woodland camouflage utilities; the opponent is depicted in desert camouflage utilities.

The Marine Corps' involvement in military operations other than war—e.g., humanitarian, peacekeeping, or evacuation missions—has greatly increased. These missions require skills that span the spectrum of conflict and support operations within a continuum of force. But the Corps' day-to-day existence also demands a responsible use of force. Nonlethal techniques are among the skills Marines use to apply a responsible use of force.

1. Unarmed Restraints and Manipulation

Marines operate within a continuum of force, particularly in support of peacekeeping or humanitarian types of missions. In these situations, Marines must act responsibly to handle a situation without resorting to deadly force. Unarmed restraints and manipulation techniques including joint manipulation, come-alongs, and takedowns can be used to control a subject without resorting to deadly force. Marines must train to become proficient in nonlethal techniques and to respond in a responsible manner. These techniques are referred to as compliance techniques, and they are applied in the third level in the continuum of force.

WARNING

During training, never apply the techniques for unarmed restraints and manipulation at full force or full speed. Use a slow and steady pressure to avoid injury.

Compliance Techniques

Compliance techniques are unarmed restraint and manipulation techniques used to physically force a subject or opponent to comply. Compliance can be achieved through the close combat techniques of-

Pain compliance using joint manipulation and pressure points. (Pain compliance is the initiation of pain to get compliance on the part of the subject.)
Come-along holds.

Principles of Joint Manipulation

Joint manipulation is used to initiate pain compliance and gain control of a subject. It involves the application of pressure on thejoints (elbow, wrist, shoulder, knee, ankle, and fingers). Pressure is applied in two ways:

In the direction in which the joint will not bend. For example, joints such as the knees

and elbows only bend in one direction and when pressure is applied in the opposite direction, pain compliance can be achieved. Beyond the point where the joint stops naturally in its range of movement (i.e., it no longer bends).

Since each joint has a breaking point, Marines should apply slow steady pressure only until pain compliance is reached. Continued pressure will break the joint and may escalate the violence of the situation.

Joint manipulation also uses the principle of off-balancing. A subject can be better controlled when he is knocked off balance.

Wristlocks

A wristlock is a joint manipulation that can be applied in a number of ways to achieve pain compliance. The wrist rotates in a number of directions and will bend in a single direction until its movement stops naturally. In a wristlock, pressure is exerted beyond that point by bending or twisting the joint. A wristlock is executed when an opponent tries to grab Marines or is successful in grabbing Marines or their equipment. A wristlock can also be performed by Marines if they wish to initiate control of an opponent.

Basic Wristlock. A basic wristlock is executed when Marines grab the opponent's left hand with the right hand. To execute the basic wristlock, Marines—

Use the right hand to grab the opponent's left hand by placing the thumb on the back of the opponent's hand so that the Marine's knuckles are facing to the left.

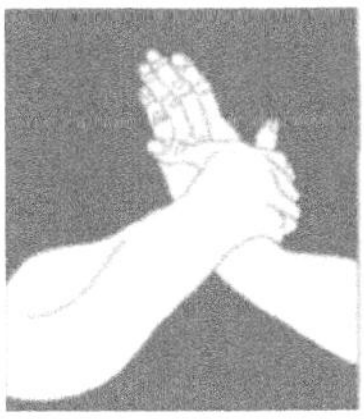

Hook the fingers across the fleshy part of the opponent's palm below the thumb. The fingers are used to anchor the hand so leverage can be applied to twist and bend the joint.

Exert downward pressure with the thumb to bend the opponent's joint. Rotate the opponent's hand to the right to twist the joint.

Step in to the opponent to keep the opponent's hand in close to the body to control him and provide more leverage on the wristlock.

Note: When executing the basic wristlock with the left hand, the Marine grabs the opponent's hand so that the Marine's knuckles are facing to the right, and then rotates and twists the opponent's hand to the left.

Use the left hand to further control the opponent.

Reverse Wristlock. A reverse wristlock is executed when Marines grab the opponent's right hand with the right hand. To execute the reverse wristlock, Marines—

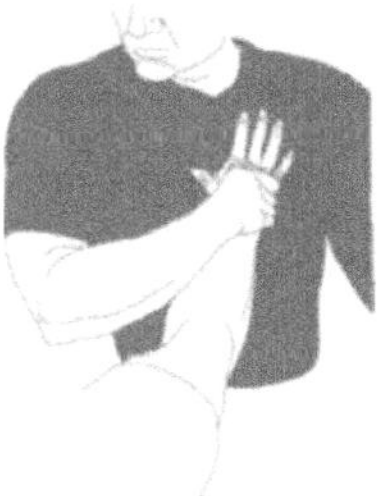

Place the right palm on the back of the opponent's right hand and wrap the fingers across the fleshy part of his palm below his little finger.

Twist the opponent's hand to the right while stepping in to place his hand against the chest. Apply downward pressure on the opponent's hand against the chest. Leave the opponent's hand on the chest to fully control the subject and to gain leverage.

Lean forward to use body weight to add additional pressure to the joint.

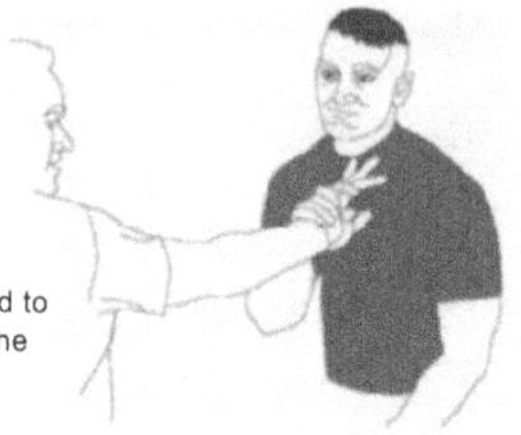

Use the left hand to further control the opponent.

Two-Handed Wristlock. Both hands can be used in the wristlock to maximize the leverage and pressure needed to bend and twist the joint. To execute the two-handed wristlock, Marines—

Place both thumbs on the back of the opponent's hand, thumbs crossed.

Hook the fingers of both hands around the fleshy part of the opponent's palm on both sides of his hand.

Step into the opponent and apply pressure downward on the back of his hand to bend the joint and rotate his wrist away from the body to twist the joint.

Enhanced Pain Compliance on Wristlock. Enhanced pain compliance techniques are applied in the third and fourth levels in the continuum of force. Additional pain be applied to a wristlock by-

Adding downward pressure to the elbow with the other hand or elbow by using the fingers to pull in on the opponent's radial nerve located on the inside of the forearm. When pressure is added to the opponent's radial nerve, his direction can be controlled.

Applying pressure against the opponent's finger joint to bend it in a direction it cannot bend (i.e., splitting the fingers).

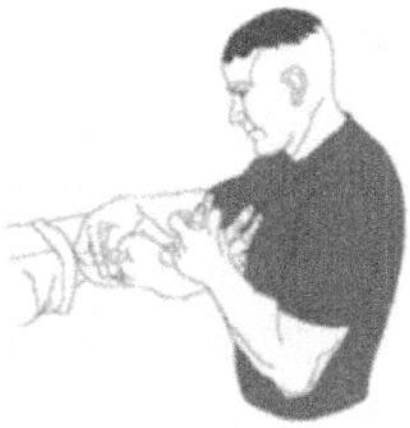

Come-Along Holds

Marines use a come-along hold to control and move an opponent.

Escort Position. A common come-along hold is the escort position. To execute the escort position, Marines—

Face the opponent. Use the left foot to step forward at a 45-degree angle. Turn to face the right side of the opponent.

Use the right hand to firmly grasp the opponent's right wrist. With the left hand, firmly grasp the opponent's right triceps.

Position the opponent's controlled arm diagonally across the torso, keeping his wrist against the right hip. The Marine should be standing to the right of and behind the opponent.

Note: This technique works well when escorting an opponent on either the right or left side. Take caution when escorting an opponent by ensuring his controlled hand is not in a position to grab the holstered weapon. The preferred escort position is from the Marine's left side, so that the opponent is kept further away from the weapon.

Wristlock Come-Along. To execute the wristlock come-along, Marines-

Use the left hand to execute a basic wristlock. Incorporate the right hand in a two-handed wristlock for more control.

Maintain pressure on the opponent's wrist with the right hand, step forward, and pivot around to stand next to the opponent.

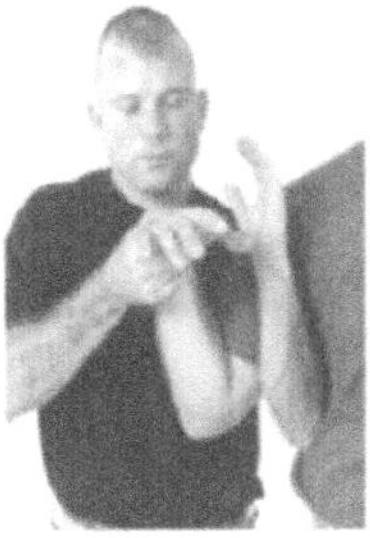

Release the left hand, quickly reach under the opponent's arm from behind, and grab his hand.

Use the left hand and apply downward pressure on the opponent's wrist.

Controlling Technique. The following controlling technique is used when an opponent grabs the Marine's wrist. To execute the technique, Marines—

Trap the opponent's hand with the palm of the other hand.

Rotate the opponent's trapped hand up and on his forearm while maintaining downward pressure on his trapped hand.

Apply downward pressure with both hands until the opponent is taken to the ground.

Armbars

An armbar is a joint manipulation in which pressure is applied on a locked elbow, just above the joint, in the direction the joint will not bend. An armbar has to be locked in quickly, but still requires a slow, steady pressure to gain compliance.

Basic Armbar. To execute a basic armbar, Marines-

Use the right hand to grab the opponent's right wrist.

Bring the left hand down on or above the opponent's elbow joint. To gain additional leverage, pivot to face the opponent.

Use the left hand to apply downward pressure on or above the opponent's elbow joint while pulling up on his wrist.

Armbar from a Wristlock. To execute an armbar from a wristlock, Marines-

Use the right hand to grab the opponent's right hand and execute a reverse wristlock.

Bring the left hand down on or above the opponent's elbow joint.

Use the left hand to apply downward pressure on or above the opponent's elbow joint while pulling up on his wrist.

Takedowns

A takedown is used to bring an opponent to the ground to further control him.

Takedown From a Wristlock Come-Along. To take the opponent to the ground from a wristlock come-along, Marines—

Use the right foot to push down on the opponent's calf or Achilles tendon.

Maintain control of the opponent's wrist and elbow and apply a slow, steady pressure to bring him to the ground.

Armbar to a Takedown. This technique is used to take a noncompliant opponent to the ground from an armbar. To execute the armbar takedown, Marines—

Use the right hand to execute a reverse wristlock.

Bring the left hand or forearm down on or above the opponent's elbow joint.

Pivot so the back is facing the opponent and, at the same time, lift the left elbow and slide the body so it is against the opponent, placing the armpit high above the opponent's elbow joint.

Lean back, placing the body weight on the opponent's arm until he complies or is taken to the ground.

Note: This technique may break the opponent's arm. Therefore, this technique should not be employed if the objective is a nonlethal takedown.

Wristlock Takedown. This technique is used to take a noncompliant opponent to the ground from a basic wristlock and to put him in a position where he can be handcuffed, if necessary. To execute the takedown, Marines-

Use the right hand to execute a basic wristlock.

Incorporate the left hand in a two-handed wristlock.

Apply downward pressure on the wristlock, pivot on the ball of the right foot, and quickly turn to the right to take the opponent to the ground.

Continue to apply pressure on the wristjoint as the opponent lands on his back with his arm straight in the air.

Slide the left foot under the opponent's back, underneath his armpit.

Apply pressure with the knee against the opponent's triceps while pulling back on his arm.

Continue applying pressure with the knee on the opponent's arm. Pivot and step around the opponent's arm to roll him on his stomach.

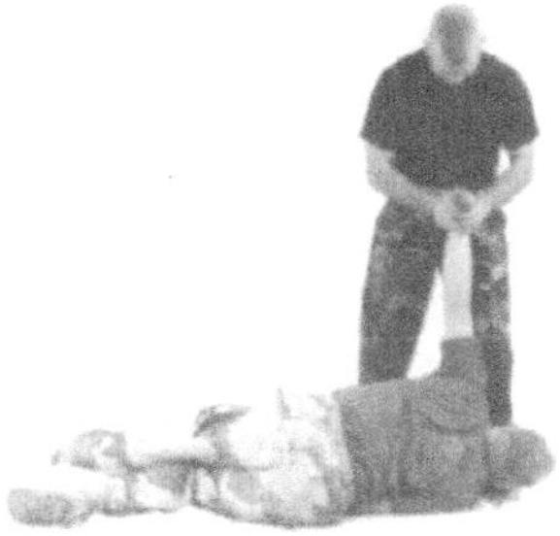

Kneel down with one knee on the opponent's back. The other knee is placed on the opponent's neck and shoulder on either side of his arm. Apply inward pressure with the knees to lock his arm in place.

Tell the opponent to put his other hand in the middle of his back. Bring the opponent's controlled hand to the center of his back.

Escort Position Takedown. This technique can be used to control a noncompliant opponent from the escort position. To execute the escort position takedown, Marines—

Lock the opponent's arm straight across the body while rotating his wrist away from the body.

Use the left hand or forearm and apply downward pressure above the opponent's elbow where the triceps meet.

Step back with the right foot and, keeping the opponent's hand controlled against the hip, pivot to the right while continuing to apply downward pressure on his arm to bring him to the ground.

Note: This technique works well when escorting a subject on the Marine's right or left side. When taking down a subject from the right side, step back and pivot to the left.

2. Nonlethal Baton

A baton or nightstick can be an effective compliance tool when used correctly. Batons can be used defensively (blocking), offensively (striking), and as a restraining device when needed. In the fourth level of the continuum of force (assaultive [bodily harm]), defensive tactics include baton or nightstick blocks and blows. Blows to the head or other bony parts of the body are considered deadly force. When deadly force is not authorized, Marines must be able to employ blocks, strikes, and restraints with the baton with the minimum of force.

Grips

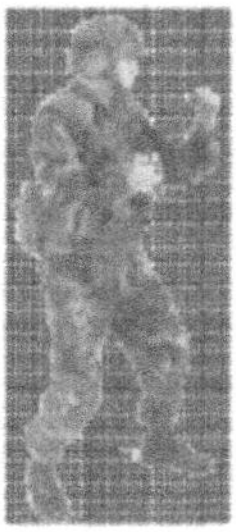

One-Handed Grip. To execute the one-handed grip, Marines—

Use the right hand to grasp the lower end of the baton, about 2 inches from the end.

Wrap the thumb and index finger around the baton so they are touching one another. The grip on the baton should be firm, but natural.

Two-Handed Grip. To execute the two-handed grip, Marines-

Use the right hand to grasp the lower end of the baton, about 2 inches from the end. Wrap the thumb and index finger around the baton so they touch one another.

Use the left hand to grasp the upper end of the baton, palm down, about 2 inches from the end. The hands should be approximately 10 to 12 inches apart.

Stance and Method of Carry

The basic warrior stance serves as the foundation for initiating nonlethal baton techniques. The method of cany provides effective defensive positions with a wide range of options to control a combative opponent.

One-Handed Carry. To execute the one-handed carry, Marines—

Grip the baton using the one-handed grip.

Elevate the baton, with the gripping hand at a level between the belt and shoulder.

Keep the left hand in the position of the basic warrior stance.

Two-Handed Carry. This carry is effective for blocks. To execute the two-handed carry, Marines-

Grip the baton using the two-handed grip.

Elevate the baton, with the left hand higher than the right hand.

Orient the weapon toward the subject.

Movement

In a nonlethal confrontation, movement may be made to create distance between Marines and the opponent or to close the gap to control him. When facing an opponent, Marines move in a 45-degree angle to either side of the opponent. Moving at a 45-degree angle is the best way to both avoid an opponent's strike and to put Marines in the best position to control the opponent.

Blocking Techniques

One-handed blocks are used when carrying the baton in the one-handed carry. The same one-handed blocks used in combative stick techniques (see page 3-6) apply in nonlethal baton engagements. The blocks include blocks for a vertical strike, a forward strike, and a reverse strike. Because the baton is often carried with two hands, there are also two-handed blocks that are effectively used from this carry. Two-handed blocks are discussed in the following subparagraphs.

High Block. Marines execute a high block to deter a downward vertical attack directed at the head and shoulders. To execute the high block, Marines—

Raise the baton up to a level even with or above the head. The baton should be in a horizontal position to block the blow. The fingers of the left hand should be open and behind the baton.

Place the baton perpendicular to the opponent's striking surface to absorb the impact of the blow.

Bend the elbows slightly to help absorb the impact of the blow. The arms should give with the strike of the blow.

Low Block. Marines execute a low block to deter an upward vertical attack directed at the abdomen, groin, or torso. The opponent's blow can be delivered by a foot, knee, or fist. To execute the low block, Marines—

Lower the baton to a level even with or below the groin. The baton should be in a horizontal position to block the blow. The fingers of the left hand should be open and behind the baton.

Place the baton perpendicular to the opponent's striking surface to absorb the impact of the blow.

Bend the elbows slightly to help absorb the impact of the blow. The arms should give with the strike of the blow.

Right Block. Marines execute a right block to deter an attack directed at the head, neck, flank, or hip. The opponent's blow can be delivered by a foot, knee, fist, or elbow. To execute the right block, Marines-

Thrust the baton in a vertical position to the right side.

Pivot to the right by stepping with the left foot and pivoting off the ball of the right foot. Rotate the hips and shoulders into the direction of the block. The fingers of the left hand should be open and behind the baton.

Place the baton perpendicular to the opponent's striking surface to absorb the impact of the blow.

Bend the elbows slightly to help absorb the impact of the blow. The arms should give with the strike of the blow.

Left Block. Marines execute a left block to deter an attack directed at the head, neck, flank, or hip. The opponent's blow can be delivered by a foot, knee, fist, or elbow. To execute the left block, Marines—

Thrust the baton in a vertical position to the left side.

Pivot to the left by stepping with the right foot and pivoting off the ball of the left foot. Rotate the hips and shoulders into the direction of the block. The fingers of the left hand should be open and behind the baton.

Place the baton perpendicular to the opponent's striking surface to absorb the impact of the blow.

Bend the elbows slightly to help absorb the impact of the blow. The arms should give with the strike of the blow.

Middle Block. Marines execute either a left or right block to deter an attack directed at the face, throat, chest, or abdomen. To execute the middle block, Marines—

Thrust the baton in a vertical position straight out in front of the body. The fingers of the left hand should be open and behind the baton.

Place the baton perpendicular to the opponent's striking surface to absorb the impact of the blow. The baton should be held with the left hand forward of the right.

Bend the elbows slightly to help absorb the impact of the blow. The arms should give with the strike of the blow.

Restraining Technique

The strong-side armlock is used to restrain an opponent who is not compliant. To execute the strong-side armlock, Marines—

Use the right hand to run the baton up under the opponent's left armpit, parallel to the ground.

Use the right foot to step forward at a 45-degree angle to the left side of the opponent. The baton is across his forearm.

Use the right hand to drive the baton forward and up so the action bends the opponent's arm behind his back. At the same time, continue moving around him to get behind him.

Place the baton on the opponent's forearm with the thumb and/or fingers and apply pressure to his forearm. At the same time, grasp the other end of the baton with the left hand.

Pull up on the low end of the baton with the right hand. At the same time, push down on the top end of the baton with the left forearm, reaching around with the left hand to grasp the opponent's biceps or shoulder.

Continue exerting downward pressure with the left forearm while pulling back on the opponent's biceps with the left hand. This places the opponent in a position where he is controlled and can be moved.

Apply pressure with the foot against the bend in the opponent's leg above his calf. This lowers the opponent to the ground rather than throwing him to the ground and risking severe injury.

Striking Target Areas

Marines must avoid striking an opponent in the head, neck, or other bony parts with the baton because this is considered deadly force and can lead to serious bodily injury or death. Instead, the legs, arms, and buttocks are target areas that are considered nonlethal.

Legs. Primary targets are the thighs and lower legs. Avoid striking the knee and ankle joints because this can cause permanent damage.

Arms. Primary targets are the upper arms. Avoid striking the shoulder and elbow and wrist joints because this can cause permanent damage.

Buttocks. Primary targets are the buttocks. Avoid striking any other part of the torso, including the chest, rib cage, spine, tail bone, and groin because strikes to these areas can cause permanent damage or death.

One-Handed Striking Techniques

One-Handed Forward Strike. A forward strike follows either a horizontal line or a downward diagonal line using a forehand stroke. To execute the one-handed forward strike, Marines-
Stand facing the opponent with the baton carried in a one-handed cany.

Place the right hand palm up, swing the baton from right to left, and make contact with the opponent.

One-Handed Reverse Strike. A reverse strike follows either a horizontal line or a downward diagonal line using a backhand stroke. To execute the one-handed reverse strike, Marines—

Stand facing the opponent with the baton carried in a one-handed carry.

Bend the right arm and cross the arm to the left side of the body. The baton should be close to or over the left shoulder.

Place the right hand palm down, swing the baton from left to right, and make contact with the opponent.

Two-Handed Striking Techniques

Two-Handed Forward Strike. This strike is an effective follow-up to a middle block or left block. To execute the two-handed forward strike, Marines—

Pull back with the left hand while driving the right hand forward toward the opponent. The baton should be horizontal to the ground. Power is generated by stepping forward with the right foot and rotating the right hip and shoulder into the strike.

Contact the opponent with the end of the baton.

Two-Handed Reverse Strike. This strike is an effective follow-up to a middle block or left block. To execute the two-handed reverse strike, Marines—

Pull back with the right hand while driving the left hand forward toward the opponent. The baton should be horizontal to the ground. Power is generated by stepping forward slightly with the left foot and rotating the left hip and shoulder into the strike.

Contact the subject with the end of the baton.

Front Jab. This strike is effective for countering a frontal attack. It can also be executed as a quick poke to keep a subject away. To execute the front jab, Marines thrust both hands forward in a quick jab. The baton is held either horizontal to the ground or at a slight downward angle.

Rear Jab. This strike is effective for countering a bear hug from the rear. To execute the rear jab, Marines thrust both hands rearward in a quick jab. The baton is held either horizontal to the ground or at a slight downward angle.

APPENDIX A

PUGIL STICK TRAINING

A pugil stick is a training device used to simulate a rifle bayonet so that effective, but safe, training can be conducted to build proficiency of rifle bayonet techniques. Pugil stick training builds on the techniques used to throw punches. Pugil stick training is the only "full contact" training provided to Marines in the Close Combat Program. Pugil stick training teaches Marines to function when faced with stress and violence, and it prepares them to deliver a blow and take a blow. It also provides them with the physical and mental skills vital to success on the battlefield.

1. Pugil Stick Training

Design

A pugil stick consists of a stick wrapped in padding at both ends that can be gripped like a rifle. The pugil stick is approximately the same weight and length of an unloaded rifle with a bayonet attached.

Holding the Pugil Stick

The pugil stick is held in the same manner as the sen ice rifle. All movements come from the basic warrior stance. To hold the pugil stick correctly, Marines—

Use the right hand to grasp the lower end of the pugil stick overhanded.
Use the left hand to grasp the upper end of the pugil stick underhanded.
Use the right forearm to lock the lower end of the pugil stick against the hip.
Orient the blade end of the pugil stick toward the opponent.

Safety Equipment

The following safety equipment must be worn during any pugil stick bout.

Groin Protection. Groin protection protects the groin from an accidental blow. It should be pulled high around the waist to protect the groin area, with the concave portion against the body.

Flak Jacket. The flak jacket provides protection to the body. It is worn completely fastened.

Neck Roll. The neck roll prevents whiplash if Marines receive a blow to the head. The neck roll further supports the head and protects the neck from blows. The neck roll is worn above the flak jacket and below the helmet. The tied end of the neck roll faces the front.

Helmet. A regulation football helmet protects the face and head. The helmet must fit snugly and the chin straps must be adjusted and snapped.

Mouthpiece. The mouthpiece is worn on the upper teeth to protect the teeth.

Gloves. Marines wear gloves to protect the hands if pugil sticks do not have gloves built-in.

Pugil Stick Screening

Prior to pugil stick training, instructors must ask participating Marines the following questions. If a Marine answers "yes" to any question, he must see the corpsman or a competent medical authority who evaluates him and determines whether or not the he can participate in the training.

Are you on light duty?
Are you restricted to running shoes by a corpsman or a doctor?
Have you fought in a pugil stick bout within the last 7 days?
Have you received a blow to the head within the last 7 days?
Have you had a concussion within the last 6 months?
Have you had dental surgery within the last 24 hours?
Do you have stitches or staples on your body?
Have you had a shoulder or head injury within the last 5 years?
Are you taking a prescription drug?
Do you have an ear infection or current sinus infection?
Have you had a broken bone within the last 6 months?

Safety Personnel

The following safety personnel are required to conduct pugil stick training:

One close combat instructor must officiate the bout.
One close combat instructor trainer, commissioned officer, or staff noncommissioned officer must be in the training area to serve as range safety officer.
One corpsman must be in the training area.

Training Area

To prevent injury, Marines train on areas with soft footing (i.e., sand or grass). Training mats are not recommended because feet can stick to the mats, prohibiting movement or causing joint injuries by twisting a knee. Bouts should not take place on a hard surface area; e.g., a flight deck or parking lot. A boxing ring may be used to conduct pugil stick bouts; ring dimensions can vary as long as there is ample room to execute the techniques in the training area.

Second Impact Syndrome

Second Impact Syndrome occurs when a second blow to the head produces a second concussion that occurs within 1 week following a previous concussion (before recovery from the first concussion). Second Impact Syndrome causes rapid brain swelling and can cause death. Therefore, there must be 7 days between pugil stick bouts to reduce the risk of severe injury resulting from Second Impact Syndrome. The 7 day separation between pugil stick bouts significantly reduces the possibility of injury, particularly in someone who may have suffered a brain injury or concussion but shows no symptoms.

Any Marine who experiences headaches or the following symptoms after training must be examined by appropriate medical personnel:

Blurred vision.
Ringing in the ears.
Dilation of the pupils.
Slurred speech.
Bleeding from ears or mouth.
Swelling in head or neck area.
Any unnatural discoloration in head or neck area.

The Marine should not be allowed to participate in pugil stick training or any other activity where a heavy blow might be sustained for a minimum of 7 days after the headache or other symptoms have subsided.

Safety Measures

The following safety measures must be followed:

Gear must be worn properly throughout training.
Contact lenses or glasses will not be worn.
False teeth will be removed from the mouth.
Nothing will be worn around the neck except the neck roll.

Competition among groups of Marines is authorized as long as it does not overshadow training objectives or compromise safety procedures.

Safety and proper techniques are paramount.

Safety is more important than competition.

2. General Rules and Regulations Governing Pugil Stick Bouts

Instructors and Support Personnel Requirements

At a minimum, one close combat instructor observes a pugil stick bout. For safety purposes, it is better to have two instructors judging a bout because each instructor can fully observe each of the fighters. The best position for observation is to the right of a fighter. This allows the instructor to see the fighter's facial expression and body movement. The instructor's position must not interfere with the fight.

In addition, one commissioned or staff noncommissioned officer (to serve as range safety officer) and one corpsman will be in the training area.

Prior to the Bout

Prior to the bout, Marines—

Are paired according to height, weight, and gender.

Wear the proper safety equipment.

Once Marines have properly donned the safety equipment, they wait for instructions from the close combat instructor. On the command of the close combat instructor, a pair of Marines will enter the ring. Once in the ring, the close combat instructor inspects each Marine for the proper safety equipment.

Note: There should he enough gear so that when two Marines are training, two other Marines can be donning safety equipment for the next bout.

During the Bout

The bout begins when the close combat instructor blows the whistle. All strikes are directed above the waist. Upon hearing a whistle blast, all fighting immediately ceases.

Stopping a Bout

There arc two reasons for stopping a bout: delivery of a scoring blow or an unsafe condition. The close combat instructor trainer, close combat instructor, or the range safety officer may stop the bout at any time an unsafe condition is observed.

Scoring Blow. A scoring blow is an offensive technique delivered to a vulnerable area of an opponent with sufficient force and precision to be considered as a disabling or killing blow. Scoring blows are not judged solely on the degree of force with which a blow is delivered, but on the accuracy and techniques employed. A scoring blow is defined as—

A straight thrust with the blade end of the weapon (red end of the pugil stick) to the opponent's face mask or throat.

A slash to the side of the opponent's helmet (below the ear) or neck with the red end of the pugil stick.

A heavy blow to the opponent's head with an authorized technique (i.e., buttstroke, smash) using the butt (black end) of the pugil stick.

When a scoring blow is delivered, the close combat instructor blows the whistle to stop the bout.

Unsafe Condition. The bout will be stopped as soon as an unsafe condition exists. An unsafe condition exists when a Marine is unable to defend himself, loses his balance and falls down on one or two knees or falls down completely, shows instability (e.g., buckling at the knees), loses muscular tension in his neck and his head snaps back or to one side, or appears disoriented. An unsafe condition also exists if a Marine lets go of one end of the pugil stick, the equipment (e.g., helmet, neck roll) falls off, or a Marine fails to use the proper techniques.

If any of these conditions occur—

The close combat instructor trainer, close combat instructor, range safety officer, or anyone supervising training stops the bout and separates the two Marines.

The corpsman evaluates the possibly injured Marine to determine if the fight can continue. The corpsman—

- Ensures that he is alert and responsive, both verbally and physically.
- Talks to him to see if he is coherent.
- Makes certain he comprehends and replies to verbal questions.
- Makes certain his speech is not slurred.
- Checks his physical signs.
- Ensures his eyes are focused and not dazed or glazed.
- Ensures his legs are not wobbly or shaky.

The pugil stick is not used as a baseball bat. The use of unauthorized techniques will result in expulsion from the bout. Only techniques taught in bayonet training are authorized. These techniques include-

Straight thrust.

Buttstroke (horizontal and vertical).

Smash.

Slash.

Parry.

Blocks (low, high, left, right).

After the Bout

A third whistle blast is used prior to identifying the Marine who delivered the killing blow.

A fourth whistle blast is used prior to demonstrating the killing technique used by the Marine who won the bout.

3. Directions for Making Pugil Sticks

Close combat instructor trainers are taught to make pugil sticks to standard specifications at school. It is recommended that all pugil sticks be made under the supervision of a close combat instructor trainer. Until the Marine Corps acquires a standardized pugil stick that units can order through the supply system, instructors use the following information and instructions to make pugil sticks locally.

Supplies

The following supplies are needed to make one pugil stick:

One 45" by 2" circular, oak, wood dowel.

Four 9" wide by 27" long foam pad pieces to make two inner handguards of double thickness.

Four 11" wide by 19" long foam pad pieces to make two outer handguards of double thickness.

Canvas material to cover inner and outer handguards.

One foam pad cut 1/4" thick by 7" wide by 13" long to make the centerguard.

One 8" wide by 14"long piece of canvas material to make a pugil stick sock.

Canvas material to make two pugil stick end caps. Ideally, two different color fabrics should be used to simulate the bayonet and buttstock ends of a rifle. Common colors for end caps are black and red. Each end cap is a different color to help the officiating certified close combat instructor determine the scoring blow during the pugil stick bout.

Two 1/4" thick by 11" wide by 54" long foam pad pieces to make two ends.

Two foam pads cut 1/4" thick by 3" wide by 14 1/2" long to make two end plug inserts.

One can of 3M Photo MountQ spray adhesive (NSN 8040-01-120-4009)to adhere foam padding to the stick. (Spray adhesive is available from self-service.)

One roll of duct tape or riggers tape (NSN 7510-00-266-5016) to protect foam pad and canvas from wear and to reinforce gluing action of spray adhesive. (Duct tape or riggers tape is available from self-service.)

One roll para-cord or 5-50 cord to tie both the inner and outer handguards together.

One roll of thread strong enough to sew canvas material together.

Sixteen grommets, one for each of the four corners of the inner and outer handguards.

One foam sleeping mat (NSN 8465-01-109-3369). Use unserviceable foam sleeping mats obtained from Defense Rcutilization and Marketing Office.

Tools

The follow ing tools are needed to make a pugil stick:

Tape measure, ruler, or yardstick.
Scissors.
Grommet machine.
Knife.
Sewing machine.

Directions

To make a pugil stick, Marines perform the following steps.

Preparing Materials

Step 1. Cut foam sleeping mat into the dimensions listed above. Cut mats to double thickness.

Step 2. Use spray adhesive to adhere the two 9" by 27" pieces of foam together. Use the spray adhesive to adhere the two 11" by 19" pieces of foam together. Let adhesive dry before using the double-thick foam for handguards.

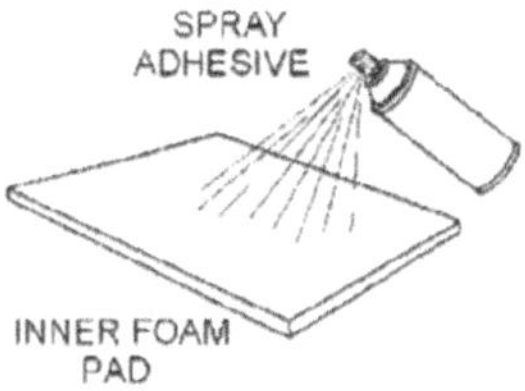

Step 3. Cut canvas material to make two coverings for the inner handguards. Cut material to make two coverings for the outer handguards. Two inner handguards and two outer handguards are required for each pugil stick.

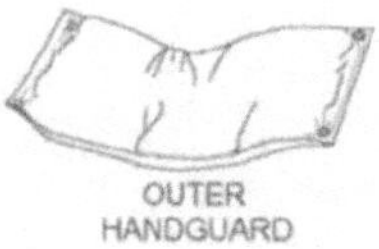

Step 4. Use a sewing machine to make a sleeve out of the 9" by 27" canvas. Sew down three sides. Insert foam pad inside the canvas sleeve and then sew down the fourth side. Use the grommet machine to create holes reinforced with metal grommets in each of the four corners. Once finished, the inner handguards arc complete.

Note: It is recommended that the Fabric Repair Shop be used to sew canvas sleeves and apply grommets.

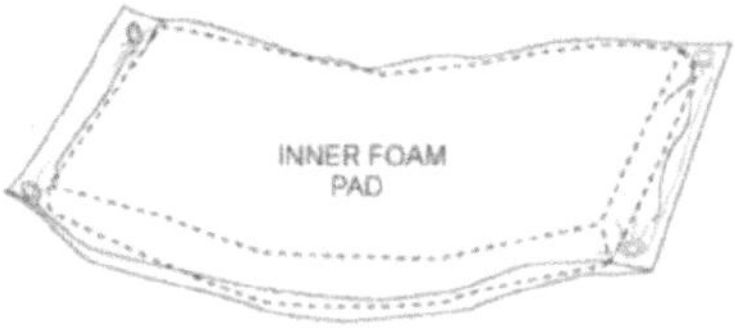

Step 5. Repeat step 4 to make the other inner handguard and both outer handguards.

Constructing Ends, End Plugs, and End Caps

Step 1. Apply spray adhesive on approximately 8 inches of one end of the dowel. This adhesive serves to glue the foam pad to the wooden dowel.

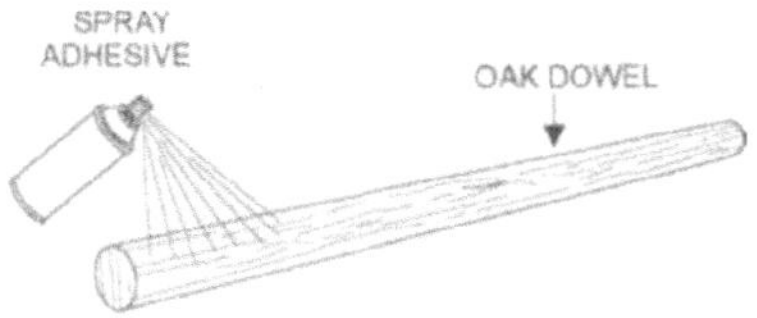

Step 2. Apply spray adhesive on one side of the 11" by 54" foam pad. This causes the foam pad to stick to itself when it is rolled around the end of the wooden dowel. Ensure 3 inches of the rolled foam pad extends past the wooden dowel for stability.

Step 3. Apply spray adhesive on one side of 3" by 14 1/2" foam pad. Roll pad into a tight cylinder.

Step 4. Apply spray adhesive on the exposed end of the dowel and fill the hole in the end with the end plug. The end plug prevents the wooden dowel from protruding beyond the edge of the rolled foam padding. This is an important safety precaution to prevent injury.

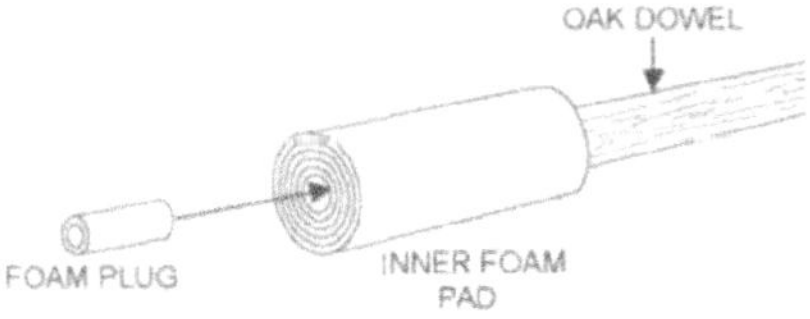

Step 5. Use duct tape or riggers tape to secure and reinforce the foam pad once the end plug is inserted in the end.

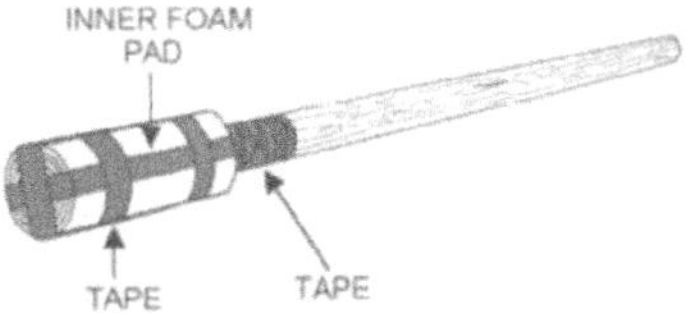

Step 6. Measure the dimensions of the ends (typically, the length of the cylinder should be 12 inches long to allow for taping). Cut and sew canvas material in the shape of a cylinder with one open end. Once complete, this becomes the end cap. The end cap is a colored sleeve that covers the rolled and taped foam pad at each end of the pugil stick.

Note: It is recommended that the Fabric Repair Shop be used to make the end cap.

Step 7. Slide one of the end caps over the rolled foam pad and tape it to the dowel.

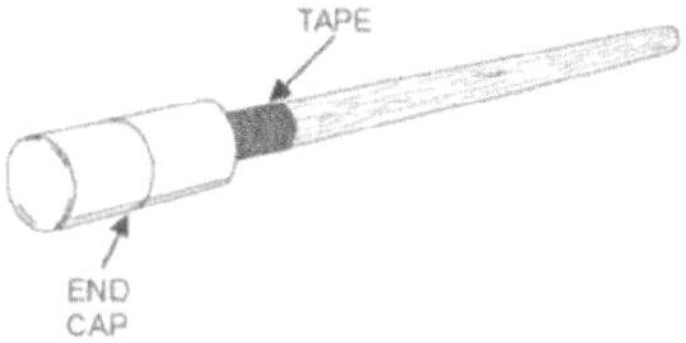

Building the Centerguard

Step 1. Spray one side of the 7" by 13" foam pad with spray adhesive and wrap it tightly around the center of the dowel.

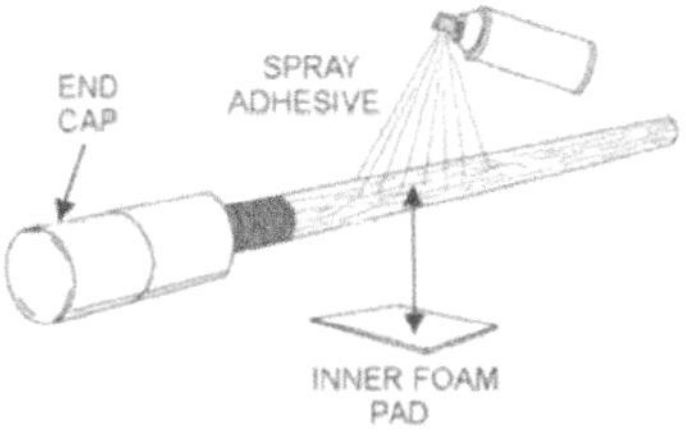

Step 2. Wrap duct tape or riggers tape around the foam pad to secure.

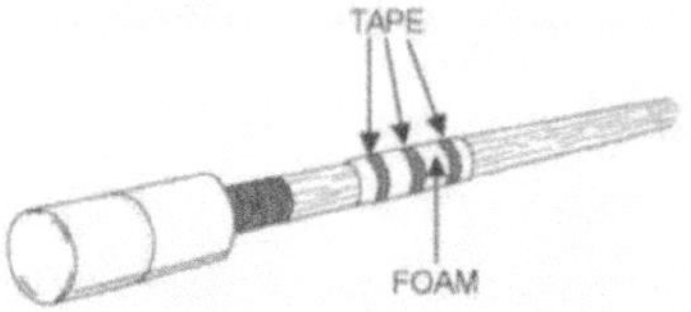

Step 3. Use a sewing machine to make a tubular pugil stick sock out of the 8" by 14" piece of canvas. Slide the pugil stick sock over the rolled foam pad and secure it with the duct tape or riggers tape. Duct tape or riggers tape protects the edges of the centerguard and reinforces the gluing effects of the spray adhesive.

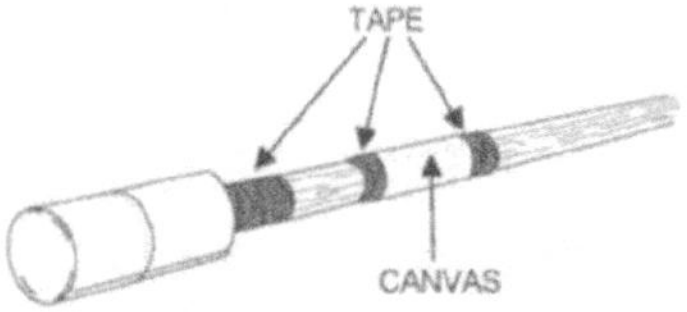

Note: It is recommended that the Fabric Repair Shop be used to make the pugil stick sock.

Finishing the Opposite End Cap

After one end cap is complete and the centerguard is finished, then repeat steps 1 through 7 in Constructing Ends, End Plugs, and End Caps for the other end of the pugil stick. Ensure that the opposite color end cap is used.

Fastening Handguards

Step 1. Place the inner handguards (longer piece) around the dowel. The inner handguard is placed on first because it protects the wrist. One inner handguard is needed for the left hand and one for the right hand.

Step 2. Fold the inner handguard around the pugil stick and run the para-cord or 5-50 cord through the metal grommets.

Step 3. Tighten down the corner ends of the inner handguards by pulling on the para-cord or 5-50 cord.

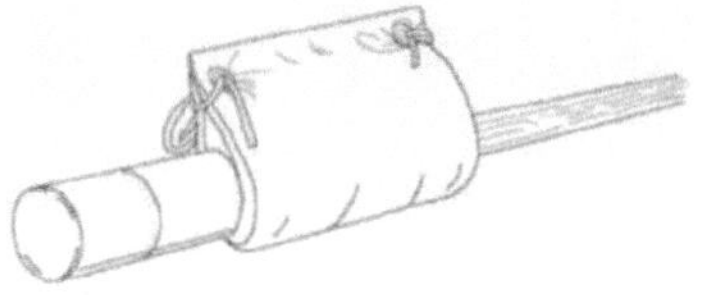

Step 4. Tie off the cord with a square knot above the metal grommets.

Step 5. Place the outer handguards (shorter piece) over the inner handguards (longer piece). Repeat steps 1-4 for the other handguard. The outer handguards provide additional protection to the knuckles and fingers and provides additional stability for the inner handguards.

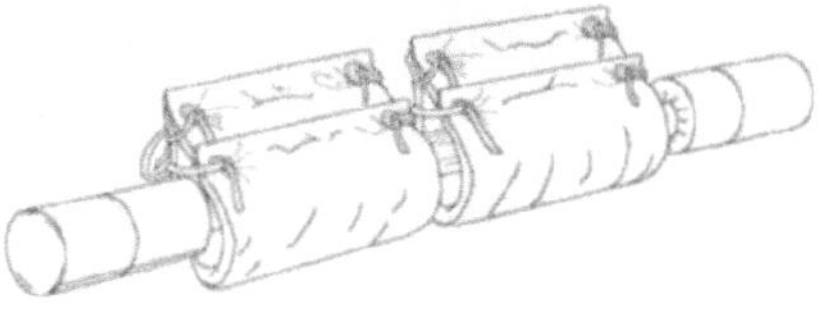

Serviceability Inspection

Prior to any training, all pugil sticks are inspected to ensure they are fit for use. If any of the following conditions exist, the pugil stick is unsafe and will not be used:

- Stick protrudes through the end caps.
- Frayed or worn edges exist on canvas surfaces.
- The stick can be felt through the pads or the padding is too soft.
- End caps or handguards are not securely attached.

(reverse blank)

APPENDIX B

SAFETY PRECAUTIONS DURING TRAINING

1. General Safety Precautions

When training close combat techniques, certain safety precautions must be adhered to to prevent injuries.

Most training should be conducted on a training area with soft footing such as a sandy or grassy area. If training mats are available, they should be used. A hard surface area is not appropriate for close combat training.

All techniques should be executed slowly at first. Marines can increase the speed of execution as they become more proficient. Marine-on-Marine training that requires contact (chokes, throws, ground fighting, and unarmed restraints and manipulation) should not be executed at full force or full speed.

If a technique is applied to the point that a Marine is uncomfortable, the Marine must "tap out." This indicates immediate release of the pressure being applied or to immediately stop the technique. The Marine "taps out" by firmly tapping his hand several times on any part of the opponent's body that will get his attention or by saying *stop*.

Second Impact Syndrome occurs when a second concussion develops within hours, days, or weeks following a prior concussion (and before recovery from the first concussion). Second Impact Syndrome causes rapid brain swelling and can cause death. Marines who experience headaches or other symptoms following training must be examined by appropriate medical personnel. These symptoms can include, but are not limited to, blurred vision, ringing in the ears, dialation of the pupils, bleeding from the ears or mouth, slurred speech, swelling in head or neck area, or any unnatural discoloration of head or neck. They should not be allowed to participate in pugil stick training or any other activity where a heavy blow might be sustained for a minimum of 7 days after the symptoms have subsided.

2. Safety Precautions for Individual Techniques

Falls

When training Marines to fall, they should progress from the ground, to a kneeling or squatting position, and then to a standing position. This ensures they are comfortable and understand the technique before progressing to executing falls from a higher profile. This instructional technique vastly reduces the risk of injury.

Strikes and Punches

When training Marines to strike and punch, they begin by executing the techniques "in the air." As they become more proficient, they execute strikes on equipment (when available) such as an air shield or a heavy bag. At no time should they be permitted to execute strikes on another student.

Chokes and Ground Fighting Chokes

When training Marines to execute chokes, they will not apply pressure to the opponent's throat during training because the trachea and windpipe can be crushed. During training, Marines should practice the proper procedures for blood chokes. They should never execute air chokes. At no time during training should a choke be applied at full force or full speed. No choke should be held for longer than 5 seconds.

Unarmed Restraints and Manipulation

When training Marines to execute unarmed restraints and manipulation techniques, they utilize slow and steady pressure. Never apply these techniques at full force or at full speed.

Pugil Stick Training

A blow to the head during training will bruise the brain. A second blow to the head can cause death. This is known as the Second Impact Syndrome. There must be 7 days between pugil stick training to prevent injury or death.

Field Manual
No. 3-19.17

Headquarters
Department of the Army
Washington, DC, 6 July 2005

Military Working Dogs

Contents

Distribution Restriction: Approved for public release; distribution is unlimited.

Figures

Tables

Preface

The last Army dog field manual (FM) was published in 1977. It reflected military working dog (MWD) doctrine developed during the Vietnam era. Although useful at that time, much of the information has since become obsolete. Today, MWD teams are employed in dynamic ways never before imagined.

Today's MWD team is a highly deployable capability that commanders have used around the world from Afghanistan to Africa and from the Balkans to Iraq. These specialized teams aid commanders in stability and support operations as well as in warfighting. Being modular and mobile makes these teams very agile. As situations dictate, MWD teams are quick to arrive and able to conduct various operations. Their versatility allows for effective transformation at all echelons among readiness for deployment and operations on the ground, through redeployment and back to readiness.

The highly aggressive dog tactics of the 1960s and 1970s are long gone. Today's MWD program effectively employs expertly trained and motivated handlers coupled with highly intelligent breeds of dogs. These teams are continuously rotating between their assigned duties and deployments worldwide to perform joint operations, multiechelon tasks, and interagency missions.

This FM addresses the current capabilities of the Military Police Working Dog Program as well as the potential for future applications. As technology and world situations change, the MWD team will continue the transformation process and give commanders the full-spectrum capabilities needed to be combat multipliers on the battlefield as well as persuasive force protection and antiterrorism assets.

Appendix A complies with current Army directives, which state that the metric system will be incorporated into all new publications.

This publication applies to the Active Army, the Army National Guard (ARNG)/the Army National Guard of the United States (ARNGUS), and the United States Army Reserve (USAR).

The proponent of this publication is United States Army Training and Doctrine Command (TRADOC). Send comments and recommendations on *Department of the Army (DA) Form 2028 (Recommended Changes to Publications and Blank Forms)* directly to Commandant, United States Army Military Police School, ATTN: ATSJ-DD, 401 MANSCEN Loop, Suite 2060, Fort Leonard Wood, Missouri 65473-8926.

Unless otherwise stated, masculine pronouns do not refer exclusively to men.

Note: For the purposes of this manual, a dog team refers to one handler and one MWD unless specified otherwise.

Chapter 1

The Military Working Dog Program

Dogs have been used for the protection of life and property since ancient times. From these beginnings, dog training and employment has been continuously refined to produce a highly sophisticated and versatile extension of the soldier's own senses. Even the most complex machines remain unable to duplicate the operational effectiveness of a properly trained MWD team.

THE ROLE OF THE MILITARY WORKING DOG

1-1. MWDs provide a valuable asset to military police, infantry, special forces, the Department of Defense (DOD), and other government agencies. The MWD's senses of sight, smell, and hearing enhance his detection capabilities and provide commanders with a physical and psychological deterrent to criminal activity. Properly trained MWDs can prevent an intruder or suspect from escaping. When necessary, the MWD provides an added dimension of physical force as an alternative to the use of deadly force. Public knowledge of MWD team capabilities provides military police and various security forces with a formidable deterrent wherever the MWD team is employed.

1-2. MWDs are key resources for use in the military police combat support role. MWDs are trained for scouting, patrolling, and performing building and area searches. Commanders can use *Table 1-1* to determine what MWD type best suits their needs. Some MWDs have also been trained to track although this is not a required area of expertise. All of these skills contribute to the successful completion of the five military police functions across the full spectrum of military operations.

Table 1-1. MWD Mission-to-Type Matrix

Mission	*MWD Type*		
	PEDD	**PNDD**	**PD**
A. Area Security Operations			
(1) Area and reconnaissance	X	X	X
(2) Screening and surveillance	X	X	X
(3) Base/air-base defense	X	X	X
(4) Cordon and search	X		
(5) Checkpoints	X	X	X
(6) Roadblocks	X	X	X
(7) Response force	X	X	X
(8) Critical site, asset, HRP security	X[1]	X	X
B. Maneuver and Mobility Support Operations			
(1) Maneuver support	X	X	X
(2) Mobility support	X	X	X
C. I/R Operations			
(1) EPW and CI	X[2]	X[2]	X[2]
(2) Evacuation	X	X	X
D. Law and Order Operations			
(1) Force protection	X	X	X
(2) Military police investigations		X	

Table 1-1. MWD Mission-to-Type Matrix (Continued)

Mission	*MWD Type*		
	PEDD	**PNDD**	**PD**
(3) Customs support	X	X	
(4) Redeployment operations	X	X	
(5) Suspicious/unattended packages	X[3]		
(6) Health and welfare inspections	X	X	
(7) Crowd control	X[4]	X[4]	X[4]
(8) Alarm responses	X	X	X
(9) Bomb threats	X		
(10) Public MWD demonstrations	X	X	X
E. PIO	X	X	X

1 PEDDs will not be used to provide security in ASPs, as the ammunition and explosives contained at the ASP will detract from the MWD's detection ability and may distract the MWD from his patrol function.

2 MWDs will not be used in a correction/detention facility to ensure custody of prisoners.

3 PEDDs can be used to search the area around a suspicious/unattended package for secondary devices. At no time will a PEDD or handler be used to search the package itself.

4 Direct confrontation with demonstrators is not recommended, but is authorized with the commander's approval when lesser means of force have been unsuccessful.

1-3. MWDs are a unique item; they are the only living item in the Army supply system. Like other highly specialized equipment, MWDs complement and enhance the capabilities of the military police. MWD teams enable the military police to perform its mission more effectively and with significant savings of manpower, time, and money.

HISTORICAL OVERVIEW OF THE MILITARY WORKING DOG

1-4. Dogs have served in active service at the sides of their handlers for decades. They have been heroes, showing bravery under fire, saving lives (often losing their own), and bringing comfort to the injured and infirmed. The first recorded American use of military dogs was during the Seminole War of 1835 and again in 1842. In Florida and Louisiana, the Army used Cuban bred bloodhounds for tracking. During the US Civil War, dogs were used as messengers, guards, and unit mascots.

1-5. The Army Quartermaster Corps began the US Armed Forces' first war dog training during World War II. By 1945, they had trained almost 10,000 war dogs for the Army, Navy, Marine Corps, and Coast Guard. Fifteen war dog platoons served overseas in World War II. Seven platoons saw service in Europe and eight in the Pacific.

1-6. MWDs were trained at Fort Carson, Colorado, organized into scout dog platoons, and used in the Korean conflict for sentry duty and support of combat patrols. In 1957, MWD training moved to Lackland Air Force base (LAFB), Texas, with the Air Force managing the program.

1-7. Throughout the Vietnam Conflict, the Military Police Corps used dogs with considerable success. Most of these were sentry dogs used to safeguard critical installations such as ports and airfields. A new dimension in canine utilization was realized when marijuana detector dog teams were trained and deployed to assist military police in suppressing illicit drug traffic. Sentry and marijuana detector dog teams were then deployed worldwide in support of military police. An important outgrowth of the conflict was the development of canine research and development efforts. These ongoing efforts were able to initiate the first steps toward developing a more intelligent and stronger military dog, training dogs to detect specific drugs and explosives, developing multiple-purpose dogs, and employing tactical dogs by electronic remote control.

1-8. In the 1990s and early 2000s, MWDs were deployed around the globe in military operations such as Just Cause, Desert Shield and Desert Storm, Uphold Democracy, and Enduring Freedom and Iraqi Freedom. These teams were effectively utilized to enhance the security of critical facilities and areas, as well as bolster force protection and antiterrorism missions, allowing commanders to use military police soldiers and other assets more effectively elsewhere. MWD operations in Haiti showed the first large-scale utilization of MWDs since Vietnam. In the Balkans, military police deployed with MWDs to provide force protection and antiterrorism support and area security throughout the US base camps. MWDs trained as explosive detector dogs conducted vehicle sweeps and building searches to ensure the safety of US soldiers. US Army MWD teams operated beside Iraqi police officials to help in law and order missions and restore the war torn country into a peaceful region.

1-9. MWDs are force multipliers. Installation commanders should include MWDs when planning for force protection and antiterrorism countermeasures. They expand the individual soldier's effectiveness in the face of those that would come against him. They provide an immense physical and psychological effect when used as a show of force in day-to-day operations. The various uses of MWDs have been effectively employed in many aspects of military police missions. MWDs are utilized effectively at gates, camps and bases, and checkpoints and for random searches for narcotics and explosive devices. MWDs are also utilized for other missions in support of combat, combat support, and combat service support units.

1-10. MWD teams are assigned to perform high-risk personnel (HRP) security missions. MWD teams perform narcotic detection missions on the Mexico and US border to assist US Customs. They have been deployed on the United States and Canada border to perform law and order operations. The canine can easily detect what his handler cannot see, hear, or smell. The MWD's controllable aggressive behavior, coupled with the physical and psychological effects he creates, makes him and his handler an impressive and unmatchable team.

ORGANIZATION AND STRUCTURE OF MILITARY WORKING DOG UNITS

1-11. The US Army MWD program is an organization with structure and responsibilities that reach from the DOD level down to the handler level. For information concerning organization and responsibilities above the local level, see *Army Regulation (AR) 190-12.*

KENNEL MASTER TEAM

1-12. The kennel master (KM) team is comprised of two noncommissioned officers (NCOs) (one sergeant first class [SFC], who is the KM and one staff sergeant [SSG], who is the plans NCO). The KM team is modular and provides technical supervision of MWD teams assigned to the military police brigade. This team may be assigned to the corps military police brigade, echelon above corps (EAC) military police brigade, or the internment and resettlement (I/R) brigade. In addition to technical supervision, this team plans and coordinates MWD operations and support requirements. The KM team is also responsible for MWD team proficiency training. This team supervises the storage and accountability of the narcotic and explosive training aids as well as all assigned equipment.

PATROL EXPLOSIVE DETECTOR DOG TEAM

1-13. The patrol explosive detector dog (PEDD) team is comprised of three NCO MWD handlers (one SSG, and two sergeants [SGTs]) and three PEDDs. This team detects explosives and contraband in support of combat, combat support, and combat service support units and the five military police functions. It provides MWD patrol support (including I/R operations) when not employed with explosive detection. This team is normally assigned to the corps military police brigade, EAC military police brigade, military police I/R brigade, or the I/R command and control (C2) detachment. This entire modular team can be deployed together as a unit with all assigned equipment or, based on the specific mission, a handler and an assigned MWD may be deployed apart from the team.

PATROL NARCOTIC DETECTOR DOG TEAM

1-14. The patrol narcotic detector dog (PNDD) team is comprised of three NCO MWD handlers (one SGT and two specialists [SPC]) and three PNDDs. This team detects narcotics and other drugs in support of combat, combat support, and combat service support units and the five military police functions. It provides MWD patrol support (including I/R operations) when not employed with narcotics detection. This team is normally assigned to the corps military police brigade, EAC military police brigade, military police I/R brigade, or the I/R C2 detachment. This entire modular team can be deployed together as a unit with all assigned equipment or, based on the specific mission, a handler and an assigned MWD can be deployed apart from the team.

INDIVIDUAL DUTIES AND RESPONSIBILITIES

1-15. There are five levels of responsibility in the MWD program. The terminal level is the program manager and the beginning level, the handler. This is not to be confused with the operational chain of command or standards of grade. These individuals have specific duties in managing and performing the MWD program.

PROGRAM MANAGER

1-16. The program manager executes major Army command (MACOM) level responsibilities for the MWD program. He is the equivalent of an operations NCO and is an appointed certification authority whose duties include—

- Conducting MWD team patrol and detector certifications.
- Requisitioning personnel and MWDs.
- Providing policy and guidelines for the MACOM.
- Forecasting annual operational budgets.
- Coordinating MWD support for the office of the Secretary of Defense, US Secret Service (USSS), US Customs Service (USCS), and Department of State missions.
- Serving as the MACOM provost marshal's (PM's) advisor for the employment of MWDs in contingency operations, force protection and antiterrorism support, and homeland defense.

KENNEL MASTER

1-17. The KM is the senior NCO in charge of the installation's MWD program, consisting of 6 to 30 handlers and respective MWDs. KM duties are the equivalent of a platoon sergeant's duties when 9 or more handlers and MWDs are assigned. When 8 or fewer handlers and MWDs are assigned, the duties are equal to those of a squad leader. In addition to traditional leadership responsibilities, KM duties include—

- Monitoring all training to ensure that MWD teams are ready for annual certifications and worldwide deployments in support of various operational missions.
- Advising the commander on the employment of MWD teams.
- Ensuring that team proficiency is maintained.
- Preparing for MACOM and veterinary inspections.
- Forecasting annual operational budgets.
- Assisting the MACOM program manager in requisitioning personnel, MWDs, and equipment.
- Managing the daily maintenance and upkeep of kennel facilities.
- Monitoring MWDs' health and welfare.
- Managing procurements and accounts for explosive and narcotic training aids.
- Advising installation and unit force protection and antiterrorism officers on the use of MWDs.
- Coordinating MWD support for the USSS, the USCS, DOD protective services units, combat support operations, the Army Recruiting Command, and other civilian agencies.

MILITARY WORKING DOG PLANS NONCOMMISSIONED OFFICER

1-18. The MWD plans NCO is part of the KM team and performs the duties of the maneuver mobility support operations (MMSO) and training NCOs. The plans NCO is selected for his ability and overall knowledge of training, mentoring, and deploying MWD assets. This NCO is responsible for—

- Developing training plans.
- Coordinating and planning the unit mission-essential task list (METL).
- Supervising the execution and tracking of the deployment plan.
- Tracking the movement of MWD teams while they are deployed.
- Ensuring that deployed soldier's issues are resolved.
- Assisting in the supervision of the installation MWD program.
- Acting as the KM when the KM is absent.
- Preparing short- and long-term training outlines and risk assessments.
- Validating and supervising proficiency training.
- Overwatching preparation for the deployment of MWD teams.

SENIOR MILITARY WORKING DOG HANDLER

1-19. The senior MWD handler is the senior NCO of the PEDD or PNDD teams. His duties are the equivalent of the squad leader of the PEDD team or the team leader of the PNDD team. In addition to traditional leadership responsibilities, he is responsible for ensuring that—

- Annual certification training is conducted.
- MWD teams are ready for deployments.
- Team proficiency is maintained.
- Handlers are proficient with their assigned MWDs.
- Coordination for MWD team support for the USSS, the USCS, combat support operations, the Army Recruiting Command, and other civilian agencies is completed.

MILITARY WORKING DOG HANDLER

1-20. The handler provides the daily care and grooming for his assigned MWD. He ensures that proficiency skills are maintained in his assigned MWD. The handler and MWD perform either PNDD operations or PEDD operations. The handler and MWD are all trained for patrol but the variant for narcotics detector or explosive detector depends on the MWD team's training. The handler also—

- Fills out training records daily.
- Grooms the assigned MWD daily.
- Trains the assigned MWD daily.
- Maintains MWD kennels daily.
- Performs additional kennels duties.

Note: For more information about additional skill identifier (ASI) Z6 standards of grade, refer to *DA Pamphlet (Pam) 611-21*.

FUNDAMENTALS FOR THE USE OF MILITARY WORKING DOGS

1-21. An MWD can be trained to respond consistently to certain sensory stimuli (sights, sounds, odors, or scents) and alert the handler. Under most circumstances, a properly trained MWD can smell, hear, and visually detect movement better than a person.

1-22. Commanders and supervisors should consider the capabilities of the MWD team when identifying and evaluating possible MWD employment. Not all MWD teams are able or equipped to perform all missions equally (see *Table 1-1*, page 1-1).

USE OF PATROL DOGS

1-23. All MWD teams are trained to perform patrol missions. One benefit of employment is the deterrence of criminal activity. The obvious presence and well-published activities of MWD teams can successfully deter trespassers, vandals, violent persons, and other would-be criminals. When the desired task is to deter unauthorized or suspicious individuals, an MWD team should be assigned to a location during day or night.

1-24. The patrol dog's (PD's) contribution to the law and order effort is most effective when the team is utilized as a walking patrol. The law enforcement duties that a PD team can perform as a walking patrol include checking or clearing buildings and patrolling parking lots, family housing areas, and troop billet areas. Mobility significantly increases the potential area of coverage. The best and most effective method is the ride awhile, walk awhile method. This concept helps cover large patrol areas, and the periodic exercise helps keep the team alert. All PD teams can be utilized for—

- Antiterrorism operations.
- Force protection and antiterrorism efforts.
- Special-reaction team (SRT) operations.
- Assistance in military police investigations.
- Walking and mobile patrols.
- Alarm responses.
- Building checks and searches.
- The identification and apprehension of individuals.
- The protection of HRPs.
- The protection of funds.
- Open-area searches.
- Civil disturbances.
- Perimeter security.
- External intrusion detection.
- Response forces.
- Access control points (ACPs) support.
- High-risk targets (HRT) support.
- Mission-essential vulnerable areas (MEVA) support.
- I/R operations.

1-25. Many of the tasks above can be performed in combat support or force protection and antiterrorism operations. In addition, PD teams can and should be utilized as a tool for area defense, providing warning and response. PD teams are effective on combat patrols and as listening posts and reconnaissance team members. The employment of PD teams should not be overlooked during I/R or enemy prisoner of war (EPW) operations. PDs are trained to apprehend suspects at or near a crime scene, stop those who may attempt to escape, and protect their handlers from harm. Some PDs may be able to locate suspects who have left the scene of a crime.

1-26. The PD team should be utilized to provide a psychological deterrent and a show of additional force protection and antiterrorism measures. PD teams should be placed on roving patrols at ACPs, vulnerable areas, and the location of HRP. These missions should increase as force protection and antiterrorism levels increase.

USE OF PATROL NARCOTIC DETECTOR DOGS

1-27. Military police commanders, planners, and supervisors have a very effective means of detecting the possession and/or use of controlled substances with the employment of a PNDD team. PNDDs are trained to recognize the scent of certain illegal substances through a program of practice and reward.

1-28. In addition to PD missions, PNDD dogs can be utilized for the following:

- Criminal Investigation Division (CID) and military police investigator drug investigations.
- USCS inspections.
- Health and welfare inspections.
- Barracks searches.
- Workplace searches.
- Random gate inspections.
- Area searches.
- Vehicle searches.
- Postal inspections.
- I/R operations.
- Aircraft and luggage searches.
- Predeployment and redeployment searches.
- Other US government agency support.

1-29. When the PNDD responds on any of the substances he is trained to detect, his handler takes the appropriate action. PNDD teams perform a valuable service by helping to rid the military community of illegal drugs and the problems associated with drug and controlled substance abuse.

USE OF PATROL EXPLOSIVE DETECTOR DOGS

1-30. PEDD teams are very useful in many searches or investigations involving explosives. The search for explosives is critically important to ensure that the community is safe and protected from would-be terrorists. PEDD teams are particularly useful if there is a need to locate one or more of these explosive items, which may have been hidden in an area.

1-31. The deterrent value and the detection capabilities of PEDD teams are a very effective countermeasure to terrorism. Public knowledge that PEDD teams are in use is a deterrent to persons who may try to use explosives illegally. The knowledge that explosives can be detected by PEDD teams at entry points or in places where explosives have been hidden can prevent a person from attempting to commit terrorist acts.

1-32. In addition to PD missions, PEDD dogs can be utilized for—

- Bomb threat incidents.
- Suspicious/unattended package incidents.
- ACP searches.
- Checkpoint searches.
- Random gate searches.
- CID and military police investigations section.
- Health and welfare inspections.
- Building and area searches.
- Postal inspections.
- Aircraft and luggage searches.
- Predeployment and redeployment searches.
- Protective service missions.
- Other US government agency support.

1-33. A well–trained PEDD team can conduct a significantly more effective search of an area or facility in a much shorter time than a number of people can. Using PEDD teams helps reduce the potential risk to persons who would otherwise have to do the search without the benefit of the dog's superior sense of smell. More detailed information concerning the employment of PD, PNDD, and PEDD teams is discussed in *Chapters 5, 6,* and *7*.

Chapter 2
Administrative Requirements

This chapter contains information that is useful in planning an MWD program or adding MWD teams to an existing program. The guidelines in this section are not meant to be absolutes for utilization, training, or maintenance. They give general guidance and information that should be considered when planning a program. Leaders will find the following information helpful for writing authorization documentation statements and for planning the local MWD program. It is important to remember that all MWDs are trained as PDs in addition to the specific type of detector dog (narcotic or explosive).

DETERMINATION OF A NEED FOR MILITARY WORKING DOGS

2-1. The installation provost marshal (PM), military police unit commander, or task force commander must take the initiative to establish the local MWD program if it does not already exist. The specific needs must be determined and the costs of the program must be justified. The decision process involves a thorough risk and crime analysis, an accurate evaluation of the requirements of the entire military police mission, and the use of the working dog yardsticks located in this chapter.

2-2. Some of the factors to consider when determining the need for MWD teams include—

- The mission of the team.
- The size of the area of responsibility of the team.
- The size of the installation or the area population.
- The number of personnel to be served by the team.
- Incident rates for all crimes.
- The present capability and the commitment of a portion of the manpower resources as handlers.
- The types of terrain on the installation or in the probable areas of deployment.
- The types of combat support missions for which MWD teams can be used.
- The number of critical facilities or areas
- The capabilities of each type of team.
- The number of HRP on the installation or in the area of responsibility.

VERIFICATION OF AUTHORIZATIONS

2-3. Initial MWD authorizations should be probationary using the following criteria:

- Adjustments must be made based on actual use data.
- A baseline ratio of 24 hours of MWD utilization to 8 hours training per week (4 hours patrol and 4 hours detection) must be maintained.
    - Utilization includes all missions (force protection/law enforcement, security, and combat support), whether patrolling, detecting drugs or explosives, or performing other functions (in combination), when a handler and dog are being employed together as a team.
    - Training includes all training activities to maintain, improve, regain, or develop dog skills such as obedience, controlled aggression, patrolling, scouting, detection of persons, and detection of drugs or explosives.
- An automatic review must be made by the respective MACOM after two years.

2-4. There is a trade-off when adjusting between the training and utilization necessary to reinforce basic skills or add new skills. For example, four hours per week may be adequate to maintain detector dog proficiency or to train in new skills, but additional training may be justified at times and training hours may be increased accordingly. This increase in training time should be accompanied by an equivalent reduction in utilization time. However, training hours should not be increased merely as a substitute for lack of utilization.

MISSION REQUIREMENTS

2-5. The mission requirements for the following support missions should be considered. These requirements should be used only as guidelines.

INSTALLATION FORCE PROTECTION AND LAW ENFORCEMENT SUPPORT

2-6. In order to provide continuous 24-hour, year-round MWD protection of the installation, nine MWDs are needed. Six of these dogs will be PEDD and three will be PNDD. This requirement is based on each MWD team being used a minimum of 24 hours per week. This does not include hours needed for training, grooming, or health care requirements. This does allow for temporary duty (TDY), handler leave, and illness.

2-7. The goal is to have one PEDD per installation available at all times for bomb threats or explosive type missions (not counted as part of the working force protection requirement). Installations with a three- or four-star headquarters require one additional PEDD authorization. This is due to the increased workload, facility criticality, facility vulnerability, location, past history of terrorist target selections, and present threat assessments.

HIGH-RISK PERSONNEL SECURITY MISSIONS

2-8. DOD tasks the Army to provide PEDD support for presidential, vice presidential, Secretary of State, and Secretary of Defense visits throughout the world. Currently, there is not a published formula to predetermine mission frequencies or requirements for these missions. On average, 40 teams inside the MACOM support the various missions above. These figures do not include travel or mission preparation time.

CUSTOMS SUPPORT

2-9. While no formula exists to quantify requirements for military customs support, this remains an ongoing mission that must be viewed as a requirement for additional MWD authorizations. MWDs are used to screen household goods shipments, privately owned vehicles (POVs), container express (CONEX) containers, military equipment, US mail, and port operations. Workload varies by installation population, location, unit missions and deployments, and the time of year.

CONTINGENCY OPERATIONS SUPPORT

2-10. Currently, approximately 50 MWD teams are supporting various contingency operations. This requirement will continue until mission completion. This figure represents 25 percent of Army MWD authorizations. Support for contingency operations has continued to rise since 1990 and will continue to do so. This increase must be factored into planning for MWD authorizations. A conservative average of 100 MWDs is needed for contingency operations on a yearly basis. This number is to ensure that MWD teams do not perform back-to-back deployments and that they receive proper proficiency training according to *AR 190-12* (4 hours per week in each specialty and 95 percent accuracy for PEDDs or 90 percent accuracy for PNDDs).

KENNEL CONSTRUCTION

2-11. When new MWD programs are started at locations where there are no facilities in place, kennel construction must be complete before the dogs arrive. The MWDs should be ordered through the military standard transportation and issue procedures (MILSTRIP) process approximately one year before the expected completion date of the kennel facility. In operations where MWD teams will be deployed for long periods of time and will be rotated, a permanent facility must be constructed within one year. Deployable field kennels may be used in the interim. The joint service MWD committee states that the DOD's design guide for MWD facilities must be referenced in planning the construction or renovation of a kennel facility.

2-12. Once the need and the justification for an MWD program are established, the PM, military police commander, or task force commander initiates a request to the installation facilities engineer to develop the kennel facility design. Before the kennel design has been finalized, the activity or installation veterinary corps officer (VCO), the physical security inspector, safety personnel, and the KM must be consulted to ensure that the minimum requirements are included in the final design. After kennel designs are completed and construction costs are estimated, the kennel project is submitted to the installation commander for approval and funding. The request for authorization of MWD teams must include the statement that the commitment has been made to fund and build kennels within one year prior to the arrival of the MWDs at the installation. Refer to *Appendix B* and *DA Pam 190-12* for details on kennel construction.

PROCUREMENT

2-13. All MWDs must be procured from the procurement section at the 341st Training Squadron, Lackland AFB, Texas. Local procurement of an MWD is prohibited. Requisition of an MWD team (one dog and one handler) requires two separate transactions. Handlers are obtained as a result of a personnel action requesting military police soldiers with the ASI Z6 in the required pay grades. MWDs are obtained as a result of a supply action. Commanders will assign an appropriately qualified handler to every MWD on hand. Assignment will be consistent with the policy of "one dog and one handler" as outlined in *AR 190-12*.

2-14. When the replacement of an MWD and not a handler is needed, requisitions are done according to *AR 700-81*. Initial requisitions for dogs should be submitted when positions for handlers are authorized. The MWDs in the Army supply system are—

- Dog, patrol: narcotic detector; national stock number (NSN) 8820-00-435-7542, line item number (LIN) D33800.
- Dog, patrol: explosives detector; NSN 8820-00-188-3880, LIN G33732.

2-15. The method of how to replace an MWD that has died, been adopted, or has undergone euthanasia is to submit a requisition in the MILSTRIP format that includes a statement that the unit or installation already has a trained handler on station. The MILSTRIP should also be submitted when the VCO has stated in writing that an MWD only has a 12-month life or work expectancy and/or is placed in deployment category IV (permanently nondeployable).

2-16. KMs should send requisitions for new or replacement dogs, in the MILSTRIP format, to their local property book office and forward a copy to their respective MACOM. The headquarters of the Air Force Security Police Agency will review all requisitions for dogs prior to processing by the 341st Training Squadron, Procurement Section.

TRAINING REQUIREMENTS

2-17. Training is needed at each installation to ensure that teams remain proficient. This realistic and challenging training should be scenario driven and relate to all of the possible operations in which MWD teams will operate according to the unit METL. Properly trained MWD teams are better prepared to accomplish all facets of MWD missions.

Initial Training

2-18. MWD teams are initially trained at the 341st Training Squadron, Lackland AFB, Texas. This training is done at a basic level for both the handler and the dog and is not intended to prepare either the handler or the dog for immediate duty in an MWD team. The training is only intended to give a basic overview of the MWD program, the health and care of MWDs, and proper training techniques. It is the responsibility of the gaining unit to train a new handler or dog on all missions conducted by Army MWD teams. Proficiency training is continuously required to ensure that teams are reliable for use in patrol, narcotic detection, and explosive detection.

Proficiency Training

2-19. The effectiveness of MWDs trained in patrol, narcotic detection, or explosive detection depends on the continual reinforcement of the abilities of the dog team through proficiency training. This training is mandatory and if it is not conducted properly, an appointed certification authority can decertify the MWD team. Failure to meet the minimum training standard for at least the 35 days prior to a certification action will result in MWD teams not being evaluated for certification.

2-20. The minimum training standard is at least four hours of training each week in patrol and at least four hours of training each week in narcotic or explosive detection. The minimum standard of proficiency to maintain certification as a detector team is 90 percent or higher (with no more than a 10 percent false response rate) for narcotic detector dogs, and 95 percent or higher (with no more than a 10 percent false response rate) for explosive detector dogs.

2-21. Failure to maintain an average that meets or exceeds the minimum standard for three consecutive training events will result in the automatic decertification of the MWD team. Decertified teams may not be used in any way, and must undergo intensive retraining. These teams may be recertified only after retraining and a consistent demonstration to an appointed certification authority. Proficiency training should be conducted both on and off leash.

Proficiency Training of Patrol Dog Teams

2-22. The effectiveness of PD teams depends on the continual reinforcement of mandatory proficiency training. PD teams must receive a minimum of 4 hours of patrol proficiency training each week. The training environment should approximate the military police working environment as closely as possible. Training should occur at varying times throughout the day and night and on varying days of the week, including weekends.

2-23. The emphasis in training is to develop the skills of both the handler and the dog so that they complement each other and the team becomes a working unit. As a proficiency evaluation criterion, the correct performance of the required task verifies that the handler understands how to control his PD and is able to do so.

2-24. Train the PD to protect with or without command based on any perceived threat to the handler. Because of the potential of serious bodily injury in this specialty, the certification of the handler and the dog as a team is very stringent. Ensure that the proficiency training of PDs includes an emphasis on critical tasks. There are no waivers for failure to accomplish a critical task by a PD team.

Weapon Fire Training

2-25. Training on weapon firing ranges is essential for the PD to become proficient and not be deterred from attacking agitators during gunfire. The dog must not attack the handler during gunfire. The firing of all weapons assigned to the handler should be done with the MWD present whenever possible. PDs can be desensitized with the firing of many different types of weapons. This can often be accomplished by arranging for the handlers to take the dogs to weapon ranges of different units.

2-26. Determine the dog's reaction to the sound of gunfire. It may be necessary for the handler to use counterconditioning techniques until the desired proficiency is achieved. Counterconditioning techniques include starting at distances of 300 meters and slowly bringing the gunfire closer to the dog (or as safety allows, bringing the dog closer to the weapon). The goal is for the dog not to bark or show any signs of aggression when the handler fires all assigned weapons. This can be a slow process over the course of several days.

OBEDIENCE COURSE TRAINING

2-27. The obedience course exposes the PD to various obstacles (see *Figure 2-1*) that simulate walls, open windows, tunnels, ramps, or steps. The dog's exposure to these obstacles reduces the amount of time required to adapt dogs to different environments. The dog learns to negotiate each of the obstacles. When confronted with a similar obstacle in the working environment, the dog is not deterred from completing his mission. The obedience course also develops the handler's ability to control the dog's behavior both on and off leash. The obedience course is not a substitute for exercise. A dog should never be required to negotiate the obedience course until he has been warmed up by proper exercise.

Figure 2-1. Obedience Course Obstacles

PROFICIENCY TRAINING OF DETECTOR DOG TEAMS

2-28. Continuous training of detector dogs is required so they maintain proficiency in the ability to seek and find drugs or explosives, depending on their specialty. Ensure that the detector training is always documented properly. Documentation and training are critical to both maintaining team certification and to establishing probable cause for a search based on a detector dog's number of correct responses.

2-29. There are several differences in the training requirements for PNDD and PEDD. Proficiency training scenarios for both types of detector dogs must be varied to avoid conditioning the dogs to a repetitious training pattern. The following factors will be varied for each proficiency training exercise:

- The general training exercise area.
- The number of training aids planted.
- The specific locations of the training aids in the area being used for training.
- The amount of explosives or drugs used in the training aid plant.
- The type of explosives or drugs used.
- The time of day or night of the training.
- The type of training aid container.
- The type of distraction material planted with the training aid.
- The length of time the training aid is left in place before the detector dog team search.
- The person used to handle and/or plant the training aid.
- The height above or below the floor level of the training aid plants.
- The size of the room or area in which the training aids are planted.

DECOYS

2-30. The decoy is the person who role-plays the primary adversary for training and evaluating the MWD team. The decoy may be a suspect, a subject, an attacker, an agitator, a drunk, an escapee, an enemy, or any of a number of other persons an MWD team may expect to encounter while performing military police duties. A decoy may also be neutral or an ally, such as another military policeman, a supervisor, a lost juvenile, or an innocent person passing through an identification check. Use both males and females for decoys, agitators, and suspect role-players. Vary the clothing the role-players are dressed in to expose the dogs to civilian, military, and the ethnic dress of different countries, if possible.

2-31. Decoys must be taught how to handle the arm protector sleeve or wrap and other aggression tools such as the bite suit (*Figure 2-2*), the hidden wrap, and leg wraps for self–protection when the dog attacks. The sleeve or wrap used should be a hidden sleeve rather than the full, heavy padded sleeve. The hidden sleeve can be concealed under a sweater or field jacket to achieve more realistic conditions for training the MWD.

Figure 2-2. Bite Suit

2-32. KMs and plans NCOs teach decoys how to understand and use the different drives the MWD shows. Once understood, these drives or needs of the MWD can be exploited to cause the desired result in training. Decoys must learn the type of pattern to take to and from the MWD when training on aggression tasks or the bite. For safety and optimum training effect, decoys will learn how to target the MWD into the bite gear and what actions to take during a partial and full bite.

VALIDATION

2-33. The KM or plans NCO will conduct a validation prior to requesting certification. Validations will then be conducted quarterly to verify the patrol and detection accuracy rates recorded on *DA Form 2807-R (Military Working Dog Training and Utilization Record)* and/or *DA Form 3992–(R) (Narcotics or Explosives Detector Dog Training and Utilization Record)*. Additional validations will be given to any patrol or detector MWD whose demonstrated proficiency appears to differ significantly from the rating recorded on his *DA Form 2807-R* and/or *DA Form 3992–R*. The validation will verify proficiency or identify the specific patrol/detector weaknesses that need corrective training.

2-34. Validations include at least 30 trials for a narcotic detector dog and at least 40 trials for an explosive detector dog. These trials will be divided equally with each substance the dog is trained to detect. Training aids will be distributed in at least five different areas over a five–day period. These areas should include (if available) vehicles, barracks, open areas, warehouses, theaters, planes, and luggage. A written record of the evaluation and validation will be maintained on record for a minimum of two years.

CERTIFICATION OF MILITARY WORKING DOG TEAMS

2-35. *AR 190-12* defines the standards and specific requirements for all certification of MWD teams. Certification of an MWD and handler is valid for one year and is immediately nullified when—

- The dog is assigned to another handler.
- The handler and MWD have been separated or have not conducted training for thirty-five or more consecutive days.
- The team fails to meet the minimum detector standard of 95 percent accuracy (explosives) and 90 percent accuracy (narcotics) with no more than a 10 percent false response rate for three consecutive months.

SCOUTING OR PATROLLING

2-36. The PD must be able to detect a person by scent, sound, and sight (on leash only). This task is rated critical. The following tasks must be accomplished to demonstrate proficiency in scouting or patrolling:

- Daytime standards.
  - Detect and respond to the scent of a person who is hidden within 75 yards, and follow the scent to the person's location.
  - Detect and respond to a sound made by a person 100 yards away, and follow this sound to the source.
  - Visually detect and respond to a person in view 100 yards away, and pursue the person on command.
  - Demonstrate proficiency in pursuit, attack, and release.
- Nighttime standards:
  - Detect and respond to the scent of a person who is hidden within 75 yards, and follow the scent to the person's location.
  - Detect and respond to a sound made by a person 100 feet away, and follow this sound to the source.
  - Demonstrate proficiency in pursuit, attack, and release.

Note. There is no nighttime standard for sight due to the dog's natural ability to revert to other senses in the dark.

VEHICLE PATROL

2-37. The PD must be able to ride quietly and calmly inside a vehicle with the handler until provoked or commanded. This is a semicritical task.

TRACKING

2-38. Tracking is the only noncritical task on which PDs are trained. Tracking is noncritical because not all PDs have tracking potential. The local KM should know which dogs have been trained to track. PDs that demonstrate a potential for tracking are identified by the KM or plans NCO at each unit, and the dogs are then further developed in this task. Although tracking is a valuable skill for the PD team, improper performance of the task does not degrade the overall effectiveness of the PD team to perform the military police mission.

2-39. PDs selected for tracking training should be trained and employed for tracking on a tracking harness. The harness serves as a useful cue to the PD that it should focus his senses on tracking. At the start of the track, the handler will give the command "Track" and begin casting for the scent. Initially, scent pads are identified and then an item with the tracklayer's scent on it will be used. The dog is required to track and alert on any items left by the tracklayer.

2-40. The minimum level of proficiency a PD trained for tracking must attain is to be able to follow a human scent at least one hour old, for 1 mile over varied terrain, and on a course with several turns. Given suitable tracking conditions, a skilled tracker dog team can follow the natural wanderings of individuals or groups of persons for at least 3.1 miles over rugged and varied terrain and on a scent track that is at least 12 hours old. While tracking, the dog is also capable of alerting his handler to the presence of tripwires and ambushes.

2-41. Evaluation of tracking tasks must include various ages and lengths of the track. The areas used to conduct tracking training and evaluations will be rotated frequently. Tracks will be at least one-third of a mile long but no longer than one mile. The age should vary from ten minutes to one hour. Tracks will include turns, but evaluators should ensure that the person laying the tracks is not trying to deceive the dog with undue tracklaying. Articles will be placed on the track for the dog to find.

INSPECTIONS

2-42. Inspections help ensure that the MWD section is in compliance with Army standards. *Appendix C* contains an example inspection memorandum to assist the inspection officials on what areas to inspect at an MWD section.

DEPARTMENT OF THE ARMY INSPECTIONS

2-43. Personnel from headquarters, Department of the Army (HQDA), conduct annual inspections of the MACOM MWD program. These inspections are used to assist the MACOM PM and program manager in moving the MACOM MWD program forward. This inspection will be on all aspects of the MACOM MWD program (see *Appendix C).*

MAJOR ARMY COMMAND INSPECTIONS

2-44. MACOM program managers conduct annual inspections of handlers, MWDs, training, team use, team proficiency, equipment, and kennel facilities. This ensures compliance with *AR 190-12*.

PROVOST MARSHAL AND COMMANDER INSPECTIONS

2-45. PMs and/or commanders conduct monthly inspections of handlers, MWDs, training, team use, team proficiency, equipment, and kennel facilities. These inspections are conducted at least monthly on one or more areas of the kennels and quarterly on all aspects of the installation MWD section to ensure compliance with *AR 190-12*.

2-46. A written record of the monthly and quarterly inspection is maintained on file with the appropriate corrective action (both in memorandum format), stating the corrective action taken for each deficiency. This written record is maintained for one year past the inspection date.

2-47. The responsible VCO will conduct an inspection of the kennel facility at least quarterly to ensure health and welfare for both the handlers and dogs assigned to the MWD section. A copy of this inspection is maintained on file for a minimum of two years. The responsible PM and/or commander takes all actions necessary to correct any deficiencies noted by the VCO. Veterinary inspections are to be conducted monthly, if possible.

TRAINING RECORDS

2-48. There are two forms that must be used to annotate MWD training, *DA Forms 2807-R* and *3992-R*. These forms are official military documents and must be maintained daily with factual information. Many times these documents are used as court documents, so the importance of the completion and accuracy of these documents cannot be stressed enough. Example forms and the procedures for filling these forms out are in *Appendix D*.

PROBABLE CAUSE FOLDERS

2-49. Probable cause folders are one of the most important methods of proving an MWD team's reliability. Legal requirements to conduct a search of private property based on the response of an MWD do not allow the probable cause folder a margin for error. All the items contained in the probable cause folder are copies. The following original records must be kept on file at the kennel facility:

- ***Department of Defense (DD) Form 1834 (Military Working Dog Service Record).*** *DD Form 1834* must be maintained on the MWD and kept as the initial form in the probable cause folder. This form must be accurate and signed on a change of a handler. When the MWD is trained on new substances, ensure that they are listed on the *DD Form 1834*. The 341st Training Squadron at Lackland AFB, Texas, must approve new substances.
- ***Lackland AFB Form 375 (Patrol Dog Certification).*** *Lackland AFB Form 375* follows the *DD Form 1834* in the probable cause folder and lists patrol certification information for the MWD.
- ***Lackland AFB Form 375a (Detector Dog Certification).*** *Lackland AFB Form 375a* follows the *Lackland AFB 375* (if dual certified), and lists postgraduation certification information for the MWD.
- ***Air Force (AF) Form 1256 (Certificate of Training).*** A copy of the handler's *AF Form 1256* follows *Lackland AFB Form 375a*. The handler provides a copy of the *AF Form 1256* for the probably cause folder.
- **Certification Letter**. This letter attests to the fact that the appointed certification authority has witnessed a demonstration of the capabilities of the dog team. This letter, at a minimum, should list the handler and dog, the whelp, the type of MWD (PEDD or PNDD), the social security number (SSN) of the handler, and the date of certification. This letter should be maintained in the probable cause folder after the handler's graduation certificate until the next certification of the team or a change of handler.
- **Quarterly Validation and Evaluation by the KM**. This document is used to show that the KM has conducted a quarterly evaluation on the MWD team. If the KM is not available, the plans NCO can also conduct this evaluation. This document follows the certification letter in the probable cause folder.
- **Training and Utilization Records**. The last items that should be maintained in the probable cause folder are the last three months of training and utilization records for the MWD team. These records include *DA Forms 2807-R* and *3992-R*. Once an additional month has passed, the latest month will be transferred to the MWD team folder in the filing system.

TRANSPORTATION REQUIREMENTS

2-50. Every unit, installation, and activity with MWD teams should assign enough support vehicles so that all missions and kennel support activities can be properly conducted and maintained. The minimum standard for transportation is one support vehicle per two MWD teams (two dogs and two handlers). Vehicle requirements should be determined by mission requirements and the overall number of MWD teams. Some modifications are necessary so that the MWD can be seen and can provide a visual deterrence during law and order or force protection and antiterrorism missions. These modifications will maximize the mission effectiveness of the dog while providing the dog with a comfortable, supportive, and protective area inside the vehicle. This type of modification is normally a kennel crate or cage secured in the rear of the vehicle; the crate or cage should have a nonslippery surface.

2-51. MWD handlers perform complex missions. Four-by-four type passenger vehicles should be used to ensure that MWD teams can get to any mission during inclement weather or on rough terrain.

2-52. Pickup trucks are essential for the transportation of explosive training aids. *AR 190-12* lists the requirements for explosives transported in government vehicles. However, pickup trucks should not be used for transporting MWDs for training or during road patrol type duties. This vehicle offers limited protection for the MWD from inclement weather and innocent bystanders.

2-53. Warning signs stating "CAUTION—MILITARY WORKING DOGS" should be placed on all sides and the rear of any and all vehicles used to transport MWDs. The signs should also be printed in the host nation language. The only exception to this is the use of unmarked vehicles for President of the United States and Vice-President of the United States missions.

NARCOTICS AND EXPLOSIVE TRAINING AIDS

2-54. Detector dog training and evaluations require the frequent use of narcotic and explosive training aids. Due to the potential safety hazards and risks involved with these aids, it is absolutely mandatory for all MWD personnel and commanders to know and follow the regulatory guidance. MWD leaders and handlers must participate in periodic review and training on the requirements noted in *AR 190-12*. These requirements are clearly stated in *AR 190-12*.

Chapter 3
Managing Military Working Dog Operations

Managing an MWD program requires attention from the program manager, the KM, and the plans NCO. This chapter will help these individuals better understand their specific requirements and concerns regarding the scheduling of various missions.

SCHEDULING AND EMPLOYMENT

3-1. The KM or plans NCO assists and advises the commander or operations officer in scheduling and employing MWD teams. The KM and plans NCO are the most qualified to determine the duty schedules, work cycles, and hours of operation to support each mission. All MWD teams should be utilized at least 24 hours per week to perform nontraining related duties.

3-2. Scheduling must allow for a minimum of four hours of patrol proficiency training per week and four hours of detector training per week. Details concerning proficiency training requirements are discussed in *Chapter 2*. Daily scheduling should include enough time for feeding, grooming, and exercising the MWDs as well as maintaining equipment and kennel facilities.

PREDEPLOYMENT OPERATIONS

3-3. MWD leaders and planners need to ensure that handlers are prepared for all possible missions before the moment of execution. Careful planning and concurrent training is an ongoing process that involves all soldiers in every kennel section. No MWD team should ever be sent on a mission without the proper preparation.

PROGRAM MANAGERS PREDEPLOYMENT RESPONSIBILITIES

3-4. MACOM program managers play a key role in MWD team readiness. The careful management of information concerning each kennel enables leaders and planners to keep the unit in a constant state of readiness for any possible deployment or no-notice mission.

3-5. Program managers keep track of the—

- Total number of personnel at each kennel in the MACOM.
- Certification status of all teams at each kennel in the MACOM.
- Deployability status of all MWDs and handlers (see *Chapter 8*).
- Total number and location of deployed MWD teams.
- Status of kennel facilities at deployment locations.
- Status of VCO support at deployment locations.

KENNEL MASTER PREDEPLOYMENT RESPONSIBILITIES

3-6. The KM's responsibility is to ensure the daily accomplishment of all missions. The readiness of all assigned MWD teams is a process that requires continuous effort from all leaders. KMs must maintain statistics and make the necessary corrections to have all personnel and equipment available for future deployments and missions.

3-7. The KM is responsible for—

- Managing the number of personnel assigned to each kennel.
- Verifying the status of certified teams and the dates of certification.
- Maintaining the deployability status of MWDs and handlers.
- Ensuring that soldiers have soldier readiness packets prepared according to *AR 600-8-101*.
- Conducting monthly and quarterly counseling.
- Maintaining accurate and detailed training records that will give the forward deployed KM an understanding of strengths and weaknesses.
- Ensuring that MWD handler equipment for short-notice deployments is prepared at all times.
- Conducting annual 9-millimeter handgun qualification. More than one qualification should be conducted (if possible) to better train MWD handlers on the use of the 9-millimeter handgun and to ensure that handlers are always qualified.
- Ensuring that handlers meet the standards outlined in *AR 600-9* and *FM 21-20* by conducting height and weight testing and physical fitness tests.
- Ensuring that the mandatory common military training and common task tests are conducted according to *AR 350-1*.
- Training soldiers in tasks associated with deployments (for example, kennel setup, utilization, and missions).
- Ensuring that MWD handlers have received any needed driver's licenses or transportation motor pool (TMP) licenses before a deployment order arrives.
- Ensuring that soldiers have received the needed individual readiness training in preparation for deployment in a combat zone.
- Conducting mine awareness training.

Plans Noncommissioned Officer Predeployment Responsibilities

3-8. Deployment planning must include coordination with the supporting military VCO. The plans NCO must consider the expected duration of the deployment and the types of facilities anticipated at the destination.

3-9. The plans NCO is responsible for ensuring that the handlers and MWDs in the MWD team or kennel section are technically proficient. The plans NCO is best able to evaluate and recommend to the KM which MWD team is most qualified and prepared for deployments and missions. Through continuous and realistic predeployment training, the plans NCO ensures that the MWD teams are proficient in—

- Crowd control training.
- Building searches.
- Tactical movements (through obstacles and from point to point).
- Gunfire training (handler and decoy).
- Defense training.
- Controlled aggression.
- Explosive detection (area searches, bunker searches, and tactical vehicle searches).
- Narcotics detection (area searches, bunker searches, and tactical vehicle searches).

Additional Predeployment Considerations

3-10. Army mobility requirements necessitate that each MWD has a shipping crate. Shipping crates are used to transport MWDs, simplify handling and loading, and give MWDs adequate space and ventilation during transport. Shipping crates may also be used as temporary kennels at intermediate sites or at the deployment destination.

3-11. The average water requirement for a single MWD is 10 gallons a day. Enough dog food must be taken to last 90 days or until resupply can be established. MWDs must be fed the standard high-performance dog food, contracted and supplied by the General Services Administration (GSA), unless otherwise directed by the supporting VCO. MWDs are fed twice daily unless the supporting VCO directs another meal frequency.

3-12. First aid kits must accompany the MWD team on every mission. Information on the deployment destination, climate, and terrain help determine the specific contents of the first aid kits. For example, if MWDs are employed on rough terrain, a pad toughener is included in the first aid kit.

3-13. All mission-related equipment must travel and arrive with MWD teams so that they can be fully operational as quickly as possible. Any equipment that is not sent with the MWD teams should be sent by other means to arrive at the location within 60 days of the MWD team's arrival. An example equipment list is in *Appendix E*.

DEPLOYMENT CONSIDERATIONS

3-14. There are many important factors to consider when leaders and MWD planners are preparing for deployments. To ensure that every deployment results in mission success, every aspect from notification to redeployment must be considered. These aspects include, but are not limited to, field kennels, training, and operations.

FIELD KENNELS AND SUPPORT FACILITIES

3-15. A suitable kennel site must be carefully selected. Before deployment, some possible locations may be selected on a map from a leader's reconnaissance. As soon as possible after arrival, the sites should be inspected to choose the most suitable location. When selecting a site, consult a VCO (whenever possible) about possible health hazards.

3-16. Locating the kennel in a congested area may be necessary to protect it from enemy attack. A temporary screen or physical barrier may be needed to block the dog's sight so he will be able to rest.

3-17. When shipping crates are used as temporary kennels, the interior, exterior, and ground beneath the crate must be cleaned daily to prevent the accumulation of moisture and waste or insect infestation. It must be raised four to six inches off the ground to allow for adequate drainage and to reduce parasite breeding places. In hot climates, place crates under trees, a tarpaulin, or plywood to provide shade and ventilation. Spread gravel (if available) around and under the crates to allow for drainage and the easy removal of solid waste. A large tent or prefabricated building can serve as a temporary storage location for equipment, rations, and other supplies.

3-18. A second tent or building erected at the kennel site to house handlers and the KM provides the kennel with sufficient manpower for security of the kennel area. Although conditions around a temporary kennel may be primitive, a high standard of sanitation is essential to prevent diseases and parasite infections. Permanent kennel facilities must be erected within one year of the arrival of MWD teams to the location of operational use. Modular expandable MWD transport containers (NSN 961120, 961130, or an equivalent) (*Figure 3-1*, page 3-4) with integrated air conditioning, heating, and equipment storage are recommended for use as temporary kennels in field environments.

DEPLOYMENT TRAINING

3-19. As soon as possible after arriving at the deployment destination and while making sure that site security is established, deployment training must begin. Deployment training allows the MWD teams to quickly develop familiarity with their new location. This is also known as right-seat ride time. The KM or plans NCO needs to evaluate or reevaluate the degree of skill of each of the MWD teams so that the KM and NCO are fully prepared to support the deployed force. Initial evaluation and training should consist primarily of field problems such as attack, search, re-attack, and scouting and basic obedience. Training should also include the handler firing his assigned weapon while maintaining control of the MWD. Some training should be done at dusk, and then in darkness, so that the handler may see (for safety) and the MWD may adjust to the noise and flash of weapons.

Figure 3-1. Modular, Expandable MWD Transport Container

Maneuver and Mobility Support Operations

3-20. MMSO consist of measures necessary to enhance combat movement and the ability to conduct movement of friendly resources in all environments. These measures ensure that commanders receive personnel, equipment, and supplies as needed. MMSO is conducted across the full spectrum of military operations. The primary focus of military police during MMSO is to ensure the swift and uninterrupted movement of combat power and logistical support. KM teams will coordinate MWD operations with the plans and operations officer or task force commander.

Area Security Operations

3-21. When setting up listening posts or conducting combat patrols, consider the availability and suitability of PDs for the particular mission. The KM or plans NCO must be briefed early regarding the mission. This will ensure that the most suitable PD team can be selected and that the handler can prepare for the mission. The handler should check the MWD for any limiting ailments and conduct training rehearsals to become familiar with the mission. The KM or plans NCO and selected handlers must be involved in planning conferences and briefings involving the utilization of an MWD team.

3-22. In defensive operations, combat patrols are used to provide early warning, to confirm intelligence information, and to detect or deter enemy action. PD teams greatly enhance the security of reconnaissance and combat patrols.

3-23. On combat patrols, a PD works at maximum efficiency for only two or three hours. During inclement or extreme hot or cold weather, this time will vary. The KM, plans NCO, and VCO should advise the commander on the MWDs' particular capabilities. The team is most effective in uninhabited areas. If an MWD frequently alerts on friendly forces and is continuously taken off his alert, he soon loses interest and reliability. Only the best teams should be selected.

Defiles and Route Reconnaissance/Surveillance Operations

3-24. MWD teams can be used to enhance the security at defile and route reconnaissance/surveillance operations. Consider the following to ensure mission success:

- The mission at hand (the possible threat scenario).
- The size of the area to be covered.

- The type and number of MWD teams.
- The duration of the mission.
- Sustainment supplies.
- Communications.

3-25. Operational planning must ensure that the proper numbers and types of teams are requested to perform the mission. Planning considerations must allow for proper rotations and rest-to-work ratios for the MWD teams.

Checkpoints

3-26. MWD teams are often called on to enhance the effectiveness of checkpoints. After receiving the mission to perform checkpoint duty, planners and handlers need to consider the following:

- The force protection condition (FPCON) (see *Appendix F* for details).
- The type and number of teams required versus those available.
- Sustainment equipment.
- Communications.
- Available support elements.

Area Defense

3-27. Areas that cannot be covered by static defensive posts because of vegetation, terrain, or some other peculiar feature can be secured using PD teams. Prior to positioning any MWD team in a defensive posture, ensure that intelligence preparation of the battlefield (IPB) details any possible placement of friendly or enemy mines. When relieving others from established defensive positions, use range cards to familiarize the team with the area.

3-28. A PD team can detect personnel using concealed avenues of approach and provide an early warning to defenders. Each MWD team's area of responsibility should be large enough so that the team can move to take advantage of the prevailing wind directions. A fixed position should be established to provide covering fire for each MWD team in case the team must withdraw closer to the defensive perimeter after the MWD has alerted on an attempted hostile probe or penetration.

Warning and Response Procedures

3-29. Several procedures should be developed for warning friendly forces of a PD's response because circumstances may prevent the use of a radio. Also, different types of responses tailored to fit a variety of potential situations should be developed. For example, it may be desirable to have the MWD team follow the alert and locate the cause. At other times, combat intelligence may require the team to maintain the alert position until assistance arrives or to withdraw to a more advantageous position. Whatever action is taken, the handler should not release an MWD unless it is necessary to defend—

- Himself.
- Other personnel.
- Protected resources.

Internment and Resettlement Operations

3-30. I/R operations consist of those measures necessary to guard, protect, and account for people that are captured, detained, confined, or evacuated by US forces. In any military operation involving US forces, accountability and the safe and humane treatment of detainees is essential. US policy demands that all persons who are captured, interned, evacuated, or held by US forces are treated humanely. This policy applies from the moment detainees become the responsibility of US forces and continues until the time they are released or repatriated. (See *ARs 190-8* and *190-47* and *FMs 3-19.40* and *27-10*.)

3-31. MWDs offer a psychological and actual deterrent against physical threats presented by personnel housed in an I/R facility. They cannot be used as security measures against US military prisoners. MWDs reinforce security measures against penetration and attack by small enemy forces that may be operating in the area. MWDs also provide a positive, effective alternative to using firearms when preventing disturbances or escapes by personnel.

3-32. MWD handlers will not use their MWDs to guard prisoners inside prisons or detainee holding facilities and confinement facilities. In addition, MWD handlers will not use their MWDs to degrade, torture, injure, or mistreat EPWs, detained personnel, civilian internees (CIs), or other detainees in US custody.

Note: MWDs are not used during interrogation operations.

Demonstrations

3-33. Conduct periodic demonstrations in full view of personnel housed in an I/R facility to increase the psychological deterrent of an MWD. Emphasize how easily and quickly an MWD can overtake a fleeing individual. Highlight the MWD's ability to attack and overcome physical resistance. Demonstrate an MWD's scouting ability. To ensure a successful demonstration, use only the most qualified MWD teams.

Perimeter Security

3-34. Use an MWD team as an addition to perimeter security by making periodic, unscheduled patrols around the perimeter fence during periods of darkness. During inclement weather, a temporary blackout or an electrical failure increases the number and frequency of patrols.

WARNING

Never set patterns or establish routines when conducting security patrols. Enemy observing security measures can exploit set times by attacking when the patrol ends or before it begins.

Narcotic and/or Explosives Detection

3-35. Walk an MWD team through living areas to search for contraband. PEDD teams can be utilized to detect explosive devices and components in the I/R facility and PNDD teams can be utilized to detect narcotics.

Work Details

3-36. Position an MWD team between the work detail and the area offering the greatest avenue of escape. MWD teams provide a valuable addition to work detail guards, particularly those employed in areas offering the greatest potential for escape. However, MWD teams are deterrents and must not be used as guards.

Police Intelligence Operations

3-37. *AR 525-13* directs all military law enforcement to develop foreign and domestic criminal intelligence. MWD teams are an integral part of the police intelligence operations (PIO) process. PIO are conducted in conjunction with all functions in which the MWD teams perform. For example, a PEDD team working at a checkpoint to detect explosives will also be gathering intelligence. This holds true for all MWD functions.

3-38. Leaders and planners of MWD operations must develop procedures to ensure that the MWD teams gather and submit all intelligence in a timely manner. Standardized reports are an effective means that allow for the accurate and detailed submission of intelligence. Contact the supporting intelligence section (S2), provost marshal office (PMO), or CID office to ensure that all PIO coordination is in place prior to performing MWD functions.

LAW AND ORDER OPERATIONS

3-39. Law and order operations are those that MWD teams perform to enforce laws, orders, and regulations and to maintain discipline and safety. Law and order operations can be performed in both combat and peacetime environments

RANDOM ANTITERRORISM MEASURES PROGRAM

3-40. The random antiterrorism measures program (RAMP) is designed to be used at all FPCON levels to increase physical security awareness throughout an installation. RAMP involves the random use of measures, identified in *AR 525-13* and *TC 19-210*, for FPCON levels to make installation operations less predictable and therefore more secure.

3-41. The use of MWD teams during RAMP provides for more thorough detection of narcotics and/or explosives and for a psychological deterrent. An MWD presence can also increase confidence in installation security in general.

3-42. KMs coordinate with the PMO or S3 to ensure that RAMP checks are successful. Coordination will include—

- The reconnaissance of patrol areas (map reconnaissance as a minimum).
- Rehearsals and inspections before operations.
- Transportation.
- Equipment needed on site.
- A threat briefing.
- Rules of engagement.
- Use-of-force criteria.
- After-action review (AAR) requirements (see *Appendix D*).

TEMPORARY DUTY MISSIONS (UNITED STATES CUSTOMS SERVICE/HIGH-RISK PERSONNEL SECURITY MISSIONS)

3-43. USCS safeguards the internal security of the United States through border searches. Because of this, they have broad powers in conducting searches. By utilizing an MWD team, all vehicles, ships, luggage, and baggage can be checked at border clearance crossings or ports. MWD teams are called upon to support local or federal authorities in detection operations. The Joint Enforcement Military Customs Operation places MWDs at border clearance ports to conduct searches of vehicles and cargo entering the United States. The MWD team takes no law enforcement action. The MWD team only searches and reports. *AR 90-12* gives a more detailed explanation of civil support by MWD teams.

3-44. The Air Force has been designated to serve as the primary service point of contact for all very important person (VIP) missions received from the Office of the Secretary of Defense, USSS, and/or Department of State. The VIP coordination officer who (under authority from the Secretary of Defense) tasks and coordinates all missions performed by DOD PEDD team personnel, regardless of service affiliation, issues all mission tasks.

3-45. MWD leaders and planners ensure that the certification of the team selected to perform the mission is well documented and current. Other considerations in planning for USCS support include—

- The travel arrangements of the MWD team.
- Government travel credit cards.
- Lodging for the handler.

- The kennel and boarding.
- VCO support.

Military Working Dog Public Demonstrations

3-46. Public demonstrations are conducted to show the abilities of MWD teams. These demonstrations are coordinated through the KM to ensure that the appropriate team is available for the target audience. The goal of public demonstrations is to deter criminal activity and to promote public acceptance of the MWD.

3-47. At no time should demonstrations make an MWD a form of entertainment. The KM or plans NCO should ensure that the details of sensitive police training techniques are not explained to the public. Avoid lengthy or frequent public demonstrations that allow MWDs to tire easily or become bored.

Health and Welfare Demonstrations

3-48. Initially, the MWD team should provide the responsible commander with a basic demonstration of the MWD's ability. The KM or plans NCO places a training aid in an undisclosed location, and the handler enters the search area with the MWD. The handler presents and conducts a systematic search of the area. When the MWD alerts, the appropriate reward is given. This demonstrates the MWD team's proficiency.

3-49. After the demonstration has been conducted, the unit should ensure that the area being searched is free of soldiers and that all rooms are unlocked. MWD teams are a tool the commander uses to find any illicit narcotic or explosive devices. It is the responsibility of the unit to properly search the room after the MWD has cleared the room. If the MWD team responds, military police personnel are called to continue the search. The MWD handler conducting the search should not search for the illicit narcotics or explosives. Unit commanders should also ensure that no personnel enter the building or area until the search is complete. This is usually accomplished by placing NCOs at various entranceways to stop personnel from entering the area.

Competitive Events

3-50. MWD teams should be encouraged and allowed to participate in competitive events, seminars, or conferences conducted by outside agencies or police canine organizations. Exposure to competitive events can challenge MWD teams to develop skills or learn advanced employment and training techniques. These events may require new tasks or methods that the handler can use to identify and apply to military missions. In this way, the team's proficiency and effectiveness are enhanced.

Recovery Operations

3-51. After every deployment or extended off-site mission, MWD leaders need to ensure that vital actions are conducted upon the return of the MWD team. These actions include—

- Conducting an AAR after every MWD deployment or extended mission to assist in capturing lessons learned. See *Appendix G* for detailed information on AARs.
- Conducting veterinary inspections.
- Returning any explosive or narcotic training aids.
- Updating training requirements and records.
- Recertifying teams.

Chapter 4
Legal Considerations

The use of MWDs has the same legal considerations as the use of a military police soldier. Each situation and location may be subject to various Status of Forces Agreements (SOFAs), general orders, and other legal restrictions. The following chapter addresses some fundamental areas. It is important to consult with the Staff Judge Advocate whenever legal questions pertaining to an MWD arise.

USE OF FORCE

4-1. Procedures for the use of force are found in *AR 190–14*. This regulation states that personnel engaged in law enforcement or security duties will use the minimum amount of force necessary to control the situation. This regulation further details the levels of force and encourages commanders to substitute nonlethal measures for firearms when they are considered adequate for safely performing force protection, antiterrorism, and law and order functions. The regulation states that in evaluating the degree of force required for specific situations, the following options should be considered in the order listed:

- Verbal persuasion.
- Unarmed defense techniques.
- Chemical aerosol irritant projectors (subject to host nation or local restrictions).
- The military police baton.
- MWDs.
- The presentation of a deadly force capability.
- Deadly force.

4-2. Although high in the order of measures, it is clear that the release of the MWD as a use of force is a nonlethal option to be used before deadly force is considered. It must be understood that the release of an MWD to pursue, bite, and hold a person is likely to result in an injury to the suspect. This injury can be very minor such as scratches or bruises; however, serious bodily harm is possible. For this reason, it is required by *AR 190-14* that all lesser means of force must be used first whenever possible.

4-3. The use of force differs between MWD handlers and regular military police because of the uniqueness of the MWD. Positive control of the MWD must be maintained at all times throughout all of the use-of-force measures. Often, just the presence or arrival of a handler with his MWD can control a situation. If the desired level of control is not achieved, verbal persuasion is the first degree of force to be used. If verbal persuasion is not enough force to maintain or gain control, the next measure may be considered. If the situation allows, military police may use unarmed defense techniques, spray irritants, and the baton before MWD handlers use the MWD as a measure of force.

4-4. A handler is not likely to use unarmed defense techniques when holding an MWD on leash. The handler must remain in positive control of the MWD; therefore he will not be able to get close enough to a suspect to employ unarmed defense techniques. The use of chemical aerosol irritant projectors or the baton would also hinder the handler's ability to keep positive control of the MWD. Therefore, the options are not reasonable measures of force available to the handler.

4-5. Military police and task force commanders, along with PMs, establish clear policies regarding the release of MWDs according to *ARs 190-12* and *190-14.* If the minimum amount of force necessary to control the situation requires the use of MWD force off leash, the handler must ensure that the following steps are followed:

- Ensure that the MWD has identified the same target as the handler before releasing him to bite.
- Give the warning order, "Halt, halt, halt, or I will release my dog." If assigned in a foreign country, give the order in the primary language of the host nation.
- Warn all bystanders to cease all movement.
- Call the MWD off the pursuit and regain control of him if the suspect indicates surrender or stops moving.

4-6. Use extreme caution when removing an MWD from a suspect. Regain and maintain leash control of the MWD until he has become calm enough to obey the commands "Heel" and "Stay." Do not release an MWD if the suspect is out of sight, unless an off-leash building search is authorized. Do not release MWDs in areas where children are present.

4-7. MWD teams are not used for crowd control or direct confrontation with demonstrators unless the responsible commander determines it to be absolutely necessary. When it is necessary, dogs are kept on a short leash to minimize the danger to innocent persons. Dogs are not released into a crowd. Civil disturbance contingency plans include specific criteria for the use of MWD teams that are consistent with *FM 3-19.15, AR 190-12,* and *AR 190–14.*

SEARCHES

4-8. Rules 311 and 313 through 316, *Manual for Courts-Martial (MCM),* prescribe HQDA policies concerning the conduct of searches and inspections and the disposition of property seized in conjunction with those activities. Property refers to property of the United States or of nonappropriated fund activities of an armed force of the United States wherever it is located.

4-9. Commanders and law enforcement personnel must be aware of the distinction between searches, seizures, inspections, and inventories and the provisions for the execution of these actions as outlined in the *MCM.* Knowledge of this information is necessary to ensure the legal use of a PNDD or PEDD.

4-10. Probable cause is a reasonable belief that a crime has been committed and that the person, property, or evidence sought in connection with the crime is located in the place or on the person to be searched. Probable cause must be established prior to searching any persons, property, or residence. Probable cause can be established in the following ways:

- Personal observation.
- Information obtained from a reliable informant or from the witness who observed the crime.
- Use of a detector dog team.
- Any combination of these three.

4-11. The detector dog team is considered a "reliable informant," according to Rule 315 (f)(2), *MCM.* The reliability of a team is established through the initial and annual certifications of the detector dog team by the certification authority and through proper documentation of the MWD's training records.

4-12. Certification establishes the foundation for an officer to grant a search authorization or warrant. Detector dogs are first certified after finishing their basic training. To meet the legal requirements permitting their operational use, each detector dog team must undergo an annual certification and a validation.

4-13. Each detector dog team undergoes a certification when first assigned to duty. A designated certification authority conducts the certification. The detector dog team must perform a demonstration for the person who authorizes any search. This may be any commander from company level up to the post commander. The guideline is that the person who witnesses the demonstration can only authorize a search of units he has command authority over. This person must also review the probable cause folder of the dog and must validate the team as reliable, in writing. This validation is good as long as the commander is assigned to that command.

4-14. The certification authority should witness a demonstration of the dog's capabilities. All substances the dog has been taught to respond on should be used. Additionally, the dog's capability to detect past/present odor should be demonstrated. At the completion of the demonstration, the results are documented in letter form, signed by the commander, and entered into the dog's probable cause folder. The MWD handler is responsible for preparing and maintaining the probable cause folder and the contents (see *paragraph 2-49*, page 2-9*)*. The probable cause folder accompanies the handler each time the dog is used on an actual search. Each document in the folder should be explained to the commander requesting a detector dog search. The commander with command authority should witness a simple demonstration of the dog's capabilities prior to a search. The above procedure should be followed whenever possible. This procedure establishes the dog's reliability on a continuing basis and is essential in establishing probable cause.

4-15. PEDD and PNDD teams must maintain high standards of accuracy to provide the desired level of confidence in their searches. PNDD teams must maintain a 90 percent accuracy rate for the number of trials conducted, while PEDD teams must maintain a 95 percent accuracy rate for the number of trials conducted. The installation commander authorizes the support of local and federal law enforcement agencies. *DOD Instruction 3025.12* governs the use of all PEDD teams. It governs the use of PNDD teams and ensures compliance with *Section 1385, Title 18, United States Code (USC) (The Posse Comitatus Act) (18 USC 1385 [The Posse Comitatus Act]).* Complete certification, recertification, and decertification details for narcotic and explosive detector MWD teams can be found in *AR 190-12*.

Chapter 5
Patrol Dogs and Characteristics

PDs can be used in traditional law and order functions, but commanders and planners should not overlook the tactical enhancements that a PD team offers to combat support operations. All aspects of military police functions can be enhanced by the employment of MWD teams. This chapter discusses many practical applications of the PD team.

EMPLOYMENT TECHNIQUES DURING COMBAT SUPPORT OPERATIONS

5-1. PD teams can perform many combat support operations. These teams enable the tactical commander to free up soldiers and employ their resources in other areas. This force multiplier is especially valuable in area security and force protection and antiterrorism operations.

PERIMETER SECURITY

5-2. PD teams are useful for perimeter (*Figure 5-1,* page 5-2) and distant support posts because they are usually located away from normal activity and large numbers of people. These posts also usually enclose large areas, and it would take a large number of single sentries to secure the area effectively. The large size of these areas allows the PD team to change positions to take advantage of the prevailing wind direction. These posts may be secured only during periods of advanced security and high threat. Occasional random posting of these areas is recommended, especially during periods of limited visibility (rain, snow, dust, smoke, and fog) or night conditions. Barriers and obstacles (such as fences, buildings, gullies, and streams) must also be identified and considered in security post planning. Place the PD team so that these obstacles offer the least interference to security.

5-3. Perimeter security, depending on the area to be covered, should use a two-team concept. The teams overlap each other while walking around the perimeter. PD teams should not come into direct contact with each other. Considerations include—

- The commander's intent.
- Priority intelligence requirements.
- Rules of engagement.
- Use-of-force criteria.
- The wind direction and distracters in the area (see *Chapter 8*).
- Friendly and enemy forces around the area.
- Rest areas within the perimeter to allow the MWD to relieve himself.
- Watchtowers or other security forces in the area.

5-4. The PD team should initiate the perimeter security operation by walking in a zigzag pattern around the exterior perimeter. The team should be allowed to go outside of the security area to ensure that no unwanted intruders are present.

Figure 5-1. Perimeter Security

INTRUSION DETECTION SYSTEM AUGMENTATION

5-5. Continuing development and use of Intrusion Detection Systems (IDSs) for security make it necessary to reassess the role of PD teams in the security of priority resources where IDSs are used. Although IDSs are designed to greatly reduce the probability of a successful area penetration, they are not designed to replace PD teams. IDSs cannot see beyond the outer clear zones, counter intrusions, or scout the progress of an intruder force after penetration. They are also subject to malfunction and breakdown. PD teams cannot counter all potential problems, but they can be used selectively to counter some of the IDS limitations. With the advice of the KM, security planners should develop plans that set an effective balance between sensor systems and PD teams. Some suggested functions for integrating PD teams into IDS-augmented security systems are as follows:

- PD teams may be used inside or outside critical sites or locations of HRP type personnel areas to give immediate response to areas not clearly seen by closed circuit television (CCTV) or to support security response teams in their role of alarm assessment
- PD teams may be used to scout hidden intruders from the point of penetration to the hiding location inside protected areas.
- PD teams may be posted outside critical sites during periods of increased threat to expand the protected area beyond the range of the IDS. This use is particularly desirable when there is heavy vegetation beyond the perimeter clear zone or when protected areas are located on or near installation perimeters and may, therefore, be easily accessible to the public.
- PD teams may be used as an element of the security response team or as an element of other designated response forces.
- PD teams may be used to compensate for decreased IDS effectiveness when poor visibility or other environmental conditions adversely affect the IDS equipment or when the IDS is not functioning.

Secondary Response Forces

5-6. Secondary response forces are typically used as blocking forces and for sweeps and counterattacks. PD teams can greatly improve the capabilities of the secondary response force. Some factors that must be considered are that—

- Blocking positions must be selected downwind from the PD team's position when PDs are used with blocking forces. In this type of mission, PDs should only be used to alert the forces of possible intruders.
- The secondary response force moves into the wind with the PD team in the lead position when the PDs are used for sweeps. It is imperative that security be placed on the PD teams to ensure that they are protected.
- A PD team can often pinpoint the exact location of individual intruders.

5-7. A PD team can help detect, alert, and catch hidden intruders while clearing a position that has been counterattacked successfully. PEDDs can be used to help clear the area of explosives. (See *Chapter 7*.)

Listening Posts

5-8. On a listening post, the PD team should be positioned forward of the tactical area to reduce distractions to the PD, yet close enough to maintain contact with friendly forces. In selecting a location, consider the primary detection senses of the PD. Whenever possible, PD team listening posts should be located downwind from any potential enemy position or dismount avenue of approach. Other locations force the PD to rely only on his hearing and sight for detection, reducing his effective use of scent.

Combat Patrols

5-9. In defensive operations, combat patrols are used to provide early warning, to confirm intelligence information, and to detect or deter enemy action. PD teams greatly enhance the security of reconnaissance and combat patrols.

5-10. On combat patrols, a PD works at maximum efficiency for only two or three hours. The team is most effective in uninhabited areas. If a PD frequently alerts on friendly forces and is continuously taken off his alert, he soon loses interest and reliability.

5-11. Patrol leaders and members must be briefed on actions to take when a dog handler is seriously wounded or killed. Dogs that have worked closely with patrol members and have developed a tolerance for one or more of them will usually allow one of the patrol members to return it to the kennel area. However, the dog may refuse to allow anyone near his handler, and other handlers may need to be called to help. The canine must be separated from the handler by—

- Coaxing the dog away with friendly words or food.
- Covering the dog with a poncho to immobilize it.
- Staking out the dog or leashing him.

5-12. A human will always take precedence over an animal. If no effort is effective, the dog may be destroyed. However, all efforts should be made to avoid this; an MWD is an expensive and valuable asset to replace.

5-13. If a situation arises where a canine is injured but the handler is not, the handler must be allowed to accompany the canine. If wounded or killed, an MWD should be evacuated using the same assets and should receive the same consideration as that given to a soldier under the same circumstances.

Tactical Patrols

5-14. PD teams can be used in tactical patrols to detect enemy presence, to help the patrol avoid discovery, and to locate enemy outposts. When the PD alerts, the handler should signal the patrol to halt until the cause of the alert can be identified and the patrol can proceed safely. If a firefight develops while the PD team is at the point position, the PD team should respond to fire team's directions and act as a regular member of the patrol.

5-15. Generally, the best locations for PD teams are directly in front of the patrol or on the flanks to enhance the flexibility and security of the formation. Prevailing wind direction should be used to improve the patrol's chances for early warning of enemy forces. If the wind direction is from the rear of the patrol, the PD is forced to rely on sight and sound and may not be as effective. The handler must concentrate on the PD to read his alert and will not be able to use a weapon rapidly. Therefore, members of the patrol should be assigned to protect the PD team.

5-16. PD teams can be used to enhance the capabilities of an ambush patrol. A PD team should be placed in front of the patrol to minimize distractions, yet close enough to maintain contact with other patrol members. PDs on ambush patrols must remain silent and not respond aggressively to approaching enemy forces.

Internment and Resettlement Operations

5-17. PD teams are used during I/R operations for camp perimeter security and to guard against escapes. The PD's keen senses of sight, smell, and hearing assist the handler and camp authorities in detecting unauthorized activity. A PD team's presence also acts as a deterrent to escapes. In the event of an escape, the PD is a valuable asset that can assist in the search and recovery of internees.

PD Duties and an Internment and Resettlement Camp in Iraq

On 13 April 2003, SGT Carey A. Ford and his MWD Rex responded to a riot at Camp Bucca, Iraq. It was believed that an Iraqi internee had taken a pistol belt and some personal equipment from a guard. The prisoner had taken this equipment back to his tent and the 600 other internees refused to cooperate with the guard force. SGT Ford and MWD Rex went into the I/R camp unarmed. The MWD team moved through the disruptive and unruly crowd providing a secure environment for the quick-reaction force to extract the prisoner and retrieve the stolen pistol belt and personal equipment. SGT Ford related that EPWs would not come anywhere near his MWD Rex. "We work that fear to our advantage."

Four days later, SGT Ford and Rex were called on to scout for and apprehend an escaped EPW. The fugitive had a substantial lead on them, having made his escape two hours prior to SGT Ford's notification. SGT Ford entered a possible unexploded ordnance and/or antipersonnel minefield with Rex, who was also trained to detect explosives. The MWD team was determined to locate the dangerous fugitive and prevent access to any possible weapons caches since many had been located in the area. After hours of relentless searching, SGT Ford and Rex located and apprehended the escapee. Their actions provided the psychological deterrent necessary to prevent future escape attempts.

Riot/Crowd Control

5-18. Extreme caution should be applied before committing PDs to a potential crowd control situation. Although the presence of a PD can be a powerful statement to an enemy or terrorist, it can elicit a reverse and potentially adverse effect on individuals in a riot/crowd control situation.

5-19. It is difficult to employ PDs properly during demonstrations or riots due to the high levels of confusion and excitement and the large number of potential antagonists. PDs should be trained to remain undistracted by loud noises and crowds. A crowd's reaction to a PD in a riot environment may be dependent on the location. In some areas of the world, the appearance of PDs at an angry crowd scene results in an escalation of violence. The crowd will often challenge soldiers to use the PDs as a measure of force. This is particularly true if a situation can be manufactured or provoked so as to be interpreted as an unreasonable use of force by the military. The presence of PDs can cause extreme fear (deterrence) in some cultures and regions. This fear can be advantageous and should be exploited when needed.

5-20. Committed PD teams may be used as an alternative to the use of deadly force to gain control of a riot situation. If the commander located on the ground directs the employment of PD teams for direct confrontation with demonstrators, PD teams should be held in reserve, out of the sight of the crowd. As the situation escalates, PD teams may be moved forward to within the sight of the crowd, but well away from the front lines.

5-21. When committed to the front lines in direct confrontation, PDs are allowed to bite only under the specific circumstances authorized by the responsible commander. PDs should never be released freely into a crowd. Other riot control force personnel should be positioned approximately three meters from any PD team when conducting any form of riot control. This is to prevent unintentional bite injuries to friendly, riot control force personnel that may happen to get between a PD and an instigator participating in the riot or demonstration.

5-22. PD teams must move a safe distance from the crowd to ensure the safety of the PD. Use of PD teams for direct confrontation with demonstrators is not recommended. According to the Army's use-of-force policy, the application of physical force against a person who cannot reasonably be suspected of committing an offense is not appropriate. Presence in a crowd does not necessarily constitute an offense or a situation where it would be reasonable to apply the degree of physical force of a PD. A PD would not be used unless the person is engaged in some other readily apparent criminal behavior such as theft, destruction of property, or assault.

EMPLOYMENT TECHNIQUES DURING FORCE PROTECTION AND ANTITERRORISM OPERATIONS

5-23. PD teams contribute significantly to the military police force protection and antiterrorism effort. PDs are directly responsible for executing specific security-related measures and for providing support to other government agencies that contribute to force protection.

5-24. MWD support is accomplished under the leadership of the respective command PM. Force protection is accomplished through an active role in physical security, HRP security, law enforcement, and antiterrorism efforts. Force protection and antiterrorism operations consist of those actions that prevent or mitigate hostile actions against DOD personnel (to include family members), resources, facilities, and critical information. During force protection operations, offensive and defensive measures are coordinated and synchronized to enable the joint force to perform while degrading opportunities for the enemy.

5-25. In operations, PD teams are used to enhance the detection capabilities of the combat support force and to provide a psychological deterrent to hostile intrusions. If the PD teams are properly located in a hostile environment, the PD's alert provides an initial warning to the presence of a hostile force. In past combat operations, PD teams often provided warning of attacks early enough to allow a response force time to deploy and to prevent enemy forces from reaching their objectives. PD teams have also helped clear protected areas of hostile persons, explosives, and weapons after attacks.

Crowd Control in Kosovo

During Task Force Falcon (2A) in Kosovo, LTC Cloy (the battalion commander of 1-36 INF, 1 Bde, and 1AD) requested MWD support to assist the Alpha company clear a main supply route (MSR) that had been blocked by 10,000 Kosvar Albanians. LTC Cloy had already deployed both quick-reaction forces to assist Company A because the crowd would not stay off the MSR. LTC Cloy called the battalion operations center and directed that the MWD teams be sent to the scene.

SFC Barnes, SSG Rogers, SSG Barters, and SGT McClintock responded with their dogs Queeny, Danny, Sandor, and Fedor and linked up with an infantry squad. The MWD teams and infantry squad moved dismounted to the town of Gjilani. The crowd was starting to become violent. LTC Cloy was in the middle of the crowd, standing in the center of the city at the critical MSR intersection. LTC Cloy attached the MWD teams to Company A. CPT Curtain, the company commander, told SFC Barnes to move the crowd away from his area and to establish a perimeter. The MWD teams moved into the diamond formation with the dogs on guard. Splitting the crowd in half, the MWD teams moved through the crowd and reached the city center MSR intersection in the center of the city. The handlers then used the dogs to move the crowd back out of the road. Company A and the four MWD teams established a perimeter line to keep the MSR open and traffic moving. They also apprehended troublemakers in the crowd.

Later that evening when the dog handlers entered into the DFAC, LTC Cloy stood up, gained everyone's attention, and stated, "SFC Barnes, you and your handlers were like Moses. Your dogs parted the Red Sea!"

Access Control Points

5-26. A PD team may be used to enhance force protection and antiterrorism efforts at ACPs. The PD team's function at an ACP is to deter unauthorized access, crime, and terrorism. The handler that is assigned to the ACP should use the PD so that the public can observe and hear the dog. Care must be taken to ensure that the handler constantly keeps the PD under control to prevent incident or injury to innocent personnel.

Building and Area Searches

5-27. A PD team is especially effective in checking and searching buildings such as commissaries, post exchanges, finance offices, banks, and warehouses. The PD team can check buildings visually while patrolling and can stop and dismount so that the handler can physically check doors and windows. These checks should be made with the PD on leash. To take maximum advantage of the PD's scenting ability, the handler should approach buildings from downwind.

5-28. If a building is open or forced entry is evident, PD teams may be used to track hidden intruders from the point of penetration to their location. PDs may be released to apprehend an intruder suspected of committing a serious offense when the only alternatives are escape or the use of deadly force.

5-29. Prior to searching a building, the handler will determine whether to make the search with the PD on or off leash. Some of the factors that must be considered include the following:

- Friendly forces or innocent bystanders in the area.
- The time of day or night.
- Evidence of forced entry.
- The type and size of the building or area to be searched.
- The danger to any innocent persons in the area.
- The PD's ability to work off leash.
- The threat level and IPB.

5-30. Prior to releasing the PD inside a building or enclosed area, the handler will give a clear voice warning to any persons inside the building or in the area. They will be told to come out immediately and that failure to comply with these instructions will result in the release of a trained MWD. An interpreter may be needed to assist the PD team in giving the warning in the host nation language. All persons will be warned that the PD may attack without warning and that they could receive physical injuries. The handler maintains voice control of his PD throughout the search.

> ***Note***: A warning similar to the one for off-leash searches will precede on-leash searches. The searching team should be followed by at least one other military policeman to protect the search team.

APPREHENSION OF SUBJECTS

5-31. The use of PDs to attack and hold an individual enemy or terrorist should not be overlooked. A PD can be sent into a building, confined space, or cave to find, attack, and hold an individual until the handler and friendly forces can arrive to place the subject into custody. Although disadvantages do exist (injury or death of the PD), the PD is better-suited than a human to immediately find and subdue a subject in an area where the interior structure is unknown to friendly forces.

WALKING PATROLS

5-32. The PD team's contribution to MWD team law enforcement efforts is most effective when the team is on foot. Some of the law enforcement duties that a PD team can do as a walking patrol include checking or clearing buildings and patrolling parking lots, dependent housing areas, and troop billet areas.

5-33. Since it is easily seen, a PD used during daylight in congested areas provides a good psychological deterrent to certain crimes. PDs are tolerant of people, and the presence of a large number of people does not significantly reduce the PD's effectiveness except when conducting detection missions.

5-34. Using the PD team during darkness or periods of reduced visibility and in areas where there are fewer distractions can enhance the PD's detection ability. Military police may not detect a person fleeing a crime scene at night, but a PD will normally be able to detect and locate a fleeing suspect. When necessary and appropriate, the PD is also much more effective at pursuing and holding the suspect.

5-35. Periodic use of PD teams around on–post schools, especially when school is starting and dismissing, may help deter potential vandalism, child molestation, exhibition, and illegal drug activities. PD teams can also provide effective security for communications facilities, military equipment, and command posts (CPs).

ALARM RESPONSES

5-36. In responding to an alarm condition at facilities such as clubs, finance offices, or banks, the PD team should be among the first military police on the scene. The PD team may be used to search and clear the building and immediate area and help with any apprehensions. If the PD must be used to find a suspect, other persons should attempt to avoid contaminating the area with scents or tracks that may confuse him.

VEHICLE PARKING LOTS

5-37. PD teams are effective in detecting and apprehending thieves and vandals in vehicle parking lots. The presence of the PD team may deter potential acts of theft and vandalism. The PD team can be most effective by alternating between vehicle and foot patrols of the parking lots. During hours of darkness, when there is no activity in a parking lot, the PD team should approach the lot from the downwind side. If the PD alerts, the handler should locate and challenge the suspect for identification.

5-38. If a suspect tries to escape or avoid apprehension and the handler is reasonably certain the suspect has committed or is attempting to commit a serious offense, the handler should give a clear warning to the suspect to halt. When lesser means of force have failed to stop the fleeing suspect, the PD may be released to pursue, attack, and hold. If it is not safe to release the PD, the handler may attempt to follow the suspect if the PD has the ability to track.

Chapter 6

Narcotics Detector Dogs and Characteristics

MWDs are trained to perform dual roles. In addition to the uses and characteristics of the PD, the PNDD is trained to detect the odor and presence of specific drugs. This specialized ability to detect certain illegal drugs makes the PNDD a valuable tool to help commanders and other government agencies maintain law and order. The PNDD can be employed in combat and/or force protection and antiterrorism operations worldwide.

EMPLOYMENT FOR COMBAT SUPPORT OPERATIONS

6-1. The PNDD provides commanders with unique capabilities in combat support environments. The narcotic detection capabilities of these dogs can be used to maintain the order and discipline of US soldiers as well as combatants and other persons involved in operations (EPWs, CIs, refugees, or other detained/interred persons [during time of war/conflict]). In addition to narcotic detection functions, PNDD teams should not be overlooked for their PD functions as discussed in *Chapter 5*.

INTERNMENT AND RESETTLEMENT OPERATIONS

6-2. In addition to perimeter and other security duties, PNDD teams assigned to I/R camps can be used to ensure that no illegal drugs are at the I/R camps by conducting searches of all areas. Commanders and planners should consider having all inbound supplies or equipment searched by PNDD teams. Inbound and outbound mail and vehicles should also be considered as a means for the movement of illegal drugs and, as such, subject to a search by PNDD teams.

POSTAL OPERATIONS

6-3. Postal inspections, including mail storage areas and drop boxes, can be conducted in conjunction with a military customs inspector (MCI) and according to *DOD Directive 5030.49*. The postmaster or MCI can request PNDD support from the area PM or military police brigade S3 with coordination through the KM team.

6-4. A PNDD team can be used to search mail that is received or shipped. Searches can be conducted as the mail comes off or before it goes on the truck prior to being shipped. The PNDD team can search up and down the conveyer belt as the mail is removed from or placed on the truck. If a conveyer belt is not being used, the handler can instruct postal personnel how to arrange the mail in a secure area so that the PNDD team can search it more effectively. When a PNDD team is used in the close quarters of a mailroom or delivery area, all nonessential personnel should remain clear of the area while the team is actively working.

6-5. When the PNDD responds to a parcel indicating the possible presence of illegal drugs, the CID or the military police investigations section will be notified for further action. At this point the PNDD team's actions are finished.

AIRCRAFT AND LUGGAGE SEARCHES

6-6. Before beginning the search, ensure that all personnel have exited the aircraft and all compartments are opened. The PNDD team searches the aircraft before searching any luggage or cargo. When approaching the aircraft, conduct a fresh-air perimeter search from the downwind side. Once a perimeter search has been conducted, the team boards the vessel at either the front or rear. Once the area has been cleared to search, the handler ensures that only one door is open to limit draft and wind current. A systematic search of all areas within the aircraft is conducted from the top to the bottom. The handler advises other military police on the scene where the PNDD displayed a positive response, and the augmenting military police conduct a detailed search of the area surrounding the response.

6-7. When conducting inbound baggage inspections, the items should be aligned in a sterile area in rows with at least six feet between rows. The handler conducts an on-leash search, presenting each piece of baggage. When searching cargo, every effort should be made to search the cargo before it is palletized. The cargo should be aligned in rows, with at lest six feet between rows, in a sterile area (same as baggage). The handler conducts an on-leash search, presenting cargo as required. Once the PNDD team has searched the cargo, it may be palletized while remaining in the sterile area prior to embarkation.

6-8. Conducting debarkation operations is the same as inbound operations. The host unit (to which the cargo belongs or is being delivered) provides a representative (E6 or above) and a work detail to assist the PNDD team. When pallets are removed from the aircraft, they are placed in a sterile area. Once in the sterile area, the work detail unloads the pallets and arranges the cargo, as required by the PNDD handler.

Force Protection and Antiterrorism Operations

6-9. PNDD teams are extremely useful in force protection and antiterrorism operations. Commanders and other government agencies use PNDD teams to protect the community from the harmful effects of illegal drugs. PNDD teams protect the force and surrounding communities by deterring criminal activity and enhancing law enforcement capabilities.

Health and Welfare Inspections

6-10. PNDD teams conduct health and welfare inspections when requested by the commander and coordinated through the KM, PMO, and S3. The commander contacts the KM directly to schedule a health and welfare inspection. The KM, in conjunction with the installation PM, the CID special agent in charge, and the military police investigations supervisor, briefs all levels of the command on the availability and capabilities of the PNDD teams available to them on the installation.

6-11. When scheduling the inspection, the commander should have the following information available:

- The requested time of inspection.
- The areas to be inspected.
- The number of living quarters.
- The number and size of common areas.
- Other pertinent information.

Appendix H provides a guide to assist handlers with health and welfare inspections.

6-12. Once the coordinating instructions have been established, the KM coordinates a time and location to brief the commander and demonstrate the PNDD team's capabilities. It is important that the unit requesting the health and welfare inspection practices operational security (OPSEC) and informs only those personnel who have an immediate need to know. This usually consists of the commander and first sergeant or sergeant major. The KM advises the PM and the military police investigation supervisor when the inspection is scheduled and if any additional support will be needed.

6-13. A systematic inspection of the living quarters and common areas is conducted. If the PNDD responds to a locked wall locker, the soldier assigned to that wall locker is summoned to open it. Once the wall locker has been opened, unit representatives conduct a detailed search of the wall locker and its contents.

Department of Defense Schools and Drug Abuse Resistance Education Program Inspections

6-14. KMs may receive requests from Drug Abuse Resistance Education (DARE) officers or DOD school principles to conduct a health and welfare inspection of a school. When conducting school inspections in conjunction with DARE, basic building inspection techniques apply. The primary concern when inspecting schools is the safety of the students and the secrecy of the search.

6-15. When possible, searches should be conducted in conjunction with a school activity, such as an assembly or fire drill. This will allow the handlers to conduct the search with a decreased risk of student involvement. Another option is to search the school after normal school hours or on a weekend or holiday. In any event, the use of the PNDD team on school grounds must be cleared by the school administration and the PM must be notified in advance.

Assistance to Military Police Investigations Section and/or the Drug Suppression Team

6-16. The military police investigation section or the drug suppression team (DST) may contact the KM to schedule an inspection for a number of reasons. PNDD teams may be requested to conduct an inspection of a residence or building where illegal drugs have already been found. In addition, unit commanders may request a PNDD team to inspect the quarters of a soldier who has shown a positive response on a urinalysis. The PNDD is not used to verify or test a substance that has already been found. PNDD handlers contact the military police investigations section or the DST in any circumstance that a PNDD has responded positively according to local standard operating procedures (SOPs).

Military Customs Missions

6-17. The US Army has been designated as the proponent service for the military customs program. According to current regulations, PNDD teams are used to inspect aircraft cargo areas and airframes when the military customs inspector program manager (MCIPM) directs these inspections. When conducting military customs missions, the handler should first inspect the vessel, then if deemed necessary by an MCI, continue with baggage.

6-18. When scheduling military customs missions, the MCIPM coordinates with the KM according to local SOPs. At a minimum, the MCIPM notifies the KM of—

- The requested time and date of the search.
- The number and types of vessels to be searched.
- The amount of cargo to be searched.
- The number of passengers and crew requiring the MCI's clearance.
- The determination on whether the search is a high-risk or selective search.

6-19. PNDD teams may inspect vessels such as military or civilian aircraft, boats or ships, vehicles, rail cars, or any other vehicle capable of carrying passengers and/or cargo according to the military customs program. These inspections are primarily conducted outside the continental United States. However, some missions in the continental US may be necessary for units returning from foreign theaters where the installation is their first stop within the customs territory of the United States. In addition, the military customs program may require preclearance for units or individuals deploying to areas outside the customs territory of the United States where an MCI has not yet been assigned or is not available. When conducting preclearance of mobilized or deploying soldiers' personal baggage while in formation, the MCI must ensure that the soldiers do not distract the PNDD team.

VESSELS, BAGGAGE, AND CARGO INSPECTIONS

6-20. When establishing sterile areas, the MCI coordinates with the PNDD handler and ensures that the area is of sufficient size with minimal distractions to ensure the optimal working environment for the PNDD. When scheduling working parties for embarkation or debarkation, the PNDD handler coordinates with the MCI to ensure that sufficient personnel will be present to accomplish the mission in a timely manner, but not so many soldiers that they might present a distraction to the PNDD. If an examination has been deemed necessary by the MCI, the handler presents each piece of baggage or equipment to the PNDD.

UNITED STATES CUSTOMS SERVICE MISSIONS

6-21. Missions in support of the USCS come from the MWD program manager via the KM. Handlers must have—

- Active government credit cards.
- Current passports.
- Appropriate civilian attire (or a civilian clothing allowance).
- Complete transportation requirements.

6-22. When temporarily assigned to the USCS, local USCS agents, policies, and procedures direct the PNDD team. Weapons are not authorized for handlers assigned to the USCS. The uniform of the day is determined by the USCS and is normally civilian attire.

6-23. Prior to deploying, every attempt should be made to conduct pretraining exercises with the USCS. These pretraining exercises will allow the MWD to acclimatize to larger quantities of contraband, which may be found while on the mission. The USCS will provide large training aids and determine the PNDD saturation point prior to deploying.

6-24. PNDD teams can expect to perform searches on aircraft, baggage, and tractor trailers as well as vehicle lines at US borders. When assigned to the USCS, the handler does not conduct physical searches of any kind. If the PNDD responds to an area, the handler advises a USCS agent and continues with the search, leaving the agent to conduct the actual physical search of the area.

6-25. Tractor trailers are searched in a USCS sterile area. Once the cargo has been offloaded and placed in standard lines, the PNDD team performs an exterior inspection of the vehicle before moving to all interior compartments and concluding with a systematic search of the cargo.

6-26. While assigned to a US border, the PNDD may be employed in vehicle lines. Vehicle lines are generally 10 to 20 vehicles in a row (placed bumper to bumper) that are inside a sterile area. The PNDD team travels up one side of the vehicle line and back down the other. The handler presents all productive areas. In the event that a USCS agent selects a vehicle for an individual vehicle inspection, that vehicle will be placed in a sterile area and the driver will open all the compartments of that vehicle. Once all occupants are clear of the vehicle, the PNDD will conduct a systematic inspection of the entire vehicle.

6-27. The PNDD will only be used in his narcotics detector role while deployed in support of the USCS or other federal agency. The PD capabilities of the PNDD will not be used for any reason. If the handler is instructed to use his PNDD in any method other than that of a narcotics detector dog, he must immediately notify his KM.

PNDD Support to US Customs Counterdrug Missions

In the summer of 2003, SGTs Payne and Mason from Fort Knox, KY, and SPC Johnson from Fort Sill, OK, were tasked to support a US Customs counterdrug mission along the southwest border between Mexico and the United States. After an overview and train up, the PNDD handlers and their assigned dogs went to work with the customs agents to detect illegal drugs at various points of entry into the country.

These teams became very proficient in detecting large quantities of illegal drugs. The PNDD teams detected narcotics in false floors, dashboards, gas tanks, seats, and hidden panels of many vehicles.

After three months of searching vehicles at checkpoints, the PNDD teams were responsible for detecting over 4,800 pounds of marijuana, over 95 pounds of cocaine, and other undisclosed contraband.

Random Vehicle Searches

6-28. The use of a PNDD at ACPs and installation gates can increase installation antiterrorism capabilities by providing a visual deterrent to terrorist groups and other criminal elements. The installation commander, task force commander, or base camp commander may task the PM to conduct random vehicle inspections according to local SOPs and the FPCON. When this occurs, the PM submits a schedule of random vehicle inspections to be conducted within a time frame that includes times, dates, and locations of inspections to be conducted. The schedule will be approved or disapproved and the inspection parameters will be dictated (will every vehicle be inspected or will vehicles be inspected at random according to the guidance provided) by the commander and returned to the PM. The PM then notifies the KM accordingly and provides him with a copy of the schedule with the commander's signature and inspection parameters. The PNDD's patrol capabilities may be used at ACPs to search for stowaways and personnel trying to gain access to the installation or a controlled area by hiding in delivery vehicles or other large cargo type vehicles. Extreme caution must be used when using the team in this manner. The commander must be aware that once the PNDD is given the command to search for personnel the MWDs aggression level will increase, thus increasing the chances of the PNDD biting personnel.

On-Call Drug Response

6-29. PNDD handlers will normally be placed in an on-call status. On call means the handler must remain within response time and distance according to the local SOP. When on call, PNDD handlers do not consume alcoholic beverages; however, they are in a duty position that allows them to wear appropriate civilian attire as well as leave the post according to the local SOPs.

6-30. When on call, PNDD handlers maintain positive control of a secondary communication device issued by the command. The secondary communication device may be a digital pager or cellular telephone as determined by the command. When recalled, the PNDD handler returns to the kennel in the most expedient legal manner to retrieve his PNDD. The PNDD will only be transported in an approved government owned vehicle (GOV). At no time will an MWD be transported in a POV. The handler may respond in appropriate civilian attire according to the local SOP.

6-31. KMs provide insight and guidance to the PM on technical matters of the SOP. The KM should provide recommendations as to acceptable distances which may be traveled while on call, required response times, the required secondary mode of communication, and any installation-specific items of interest, such as an alternate uniform based upon the customary civilian attire that is prominent in the area.

Chapter 7

Explosive Detector Dogs and Characteristics

In addition to the uses and characteristics of the PD, the PEDD is also trained to detect the odor and presence of explosives. This specialized ability to detect explosives makes the PEDD a valuable asset to help commanders and other government agencies maintain law and order, while enhancing safety procedures. PEDD teams can be employed in combat and/or force protection and antiterrorism operations worldwide. PEDD teams are valuable assets. The task force commander, the PM, and military police planners should consider the use of PEDD teams during combat support operations. This chapter details sound employment techniques and the capabilities of the PEDD teams that enable leaders to develop plans.

INTERNMENT AND RESETTLEMENT OPERATIONS

7-1. The Army is the DOD executive agent for all EPW/CI operations. The military police is tasked with coordinating shelter, protection, accountability, and sustainment for I/R operations. The I/R function addresses military police roles with US military prisoners, EPWs, CIs, refugees, or other detained persons (during time of war/conflict). Although the combat support military police unit initially handles EPWs/CIs, modular military police (I/R) battalions with assigned military police guard companies and supporting MWD teams are equipped and trained to perform I/R operations for sustained amounts of time (see *FM 3-19.40*).

7-2. PEDD teams may be used for searching internees' belongings or confinement cells. PEDD teams should not be used to search a person. A PEDD is trained to attack with or without command; any sudden movements the internee may make could be perceived by the PEDD as an aggressive act. Safety must remain a priority for both the handler and military personnel.

7-3. The commander can also use PEDD assets to search the property of attached military personnel when needed. This is especially helpful when new units are rotating or replacing old units. These types of searches ensure that no explosives are taken out or brought in by US forces.

POSTAL OPERATIONS

7-4. Postal inspections, including mail storage areas and drop boxes, can be conducted in conjunction with an MCI and according to *DOD Directive 5030.49*. The postmaster or MCI can request PEDD support from the area PM or military police brigade S3 with coordination through the KM team.

7-5. The PEDD team can be used to search mail that is received or shipped. Searches can be conducted as the mail comes off or before it goes on the truck prior to being shipped. The PEDD can search up and down the conveyer belt as the mail is removed from or placed on the truck. If a conveyer belt is not being used, the handler can instruct postal personnel how to arrange the mail in a secure area so that the dog can search the mail more effectively. When a PEDD team is used in the close quarters of a mailroom or delivery area, all nonessential personnel should remain clear of the area while the team is actively working.

7-6. When the dog responds to a parcel, indicating the possible presence of explosives, the area is evacuated immediately. Military police secure the area and set up a temporary CP a safe distance from the suspected explosives. Use the stand-off distance according to *Appendix I*, as applicable. Use the PEDD team to search the safe area for any possible secondary explosives. Explosive ordnance disposal (EOD) personnel are requested to dispose of the explosives or render them safe. EOD personnel may give explosives and improvised explosive device (IED) components to federal agencies for further investigation or turn over the scene to federal agencies after explosive hazards have been neutralized. The PEDD team remains at the temporary CP and resumes detection operations once EOD personnel remove the explosives.

CHECKPOINTS AND ROADBLOCKS

7-7. PEDD teams can be used at checkpoints and roadblocks to supplement the control and movement of vehicles, personnel, and materiel (*Figure 7-1*). PEDD teams are added to security forces to detect explosives and to prevent illegal actions that may aid the enemy. Using these control measures deters terrorist activities, saboteurs, and other threats.

Figure 7-1. Checkpoint Search

7-8. When using PEDD teams at hasty or deliberate checkpoints and roadblocks, a search area must be designated for vehicles and a separate area must be designated for the vehicle occupants or any dismounted personnel. The PEDD team conducts a search of the vehicle only after security forces at the checkpoint ensure that all of the occupants have exited the vehicle and that a threat to friendly forces does not exist. Security forces at the checkpoint provide overwatch while the PEDD team searches for explosives.

Note: At no time should a PEDD handler be used to search the interior of a vehicle. PEDD handlers are trained on searching techniques with a PEDD and should not be used for security force tasks.

7-9. In the event that the dog indicates the presence of explosives, the PEDD team withdraws immediately from the vehicle. The area is evacuated and EOD personnel and the chain of command are notified. The PEDD team stays behind a barrier at the safe-distance location in case the EOD team leader requests further detector assistance. (See *Appendix I* for stand-off distances.)

Note: Be sure that radio operations are stopped until the PEDD team and security forces reach a safe distance from the suspected explosives.

CORDON AND SEARCH OPERATIONS

7-10. Military police leaders and other task force commanders should consider using PEDD teams when conducting cordon and search operations. PEDDs are trained and able to greatly improve the combat force's success in uncovering hidden explosives and people.

7-11. Once the military police or other security forces have established the cordon, the PEDD teams should move in and operate with the search elements. During the search, the PEDD teams provide warning of hidden explosives or persons while the rest of the search element employs overwatch and search techniques.

7-12. The search element leader directs the actions of the team if people or explosives are detected. The PEDD handler should be prepared to advise the commander on the capabilities of the dog and actions required by the rest of the team to ensure the PEDD team's safety. It is important that all members of cordon and search teams rehearse together before conducting operations.

FORCE PROTECTION AND ANTITERRORISM OPERATIONS

7-13. PEDD support for force protection and antiterrorism operations can be requested anywhere in the world and in different circumstances. In forward deployed areas or hostile environments, PEDD teams are often requested to ensure that any suspected explosives are detected before harm can come to US Army, joint, and allied forces. In peacetime environments around the world, PEDD teams perform the detection operations much the same as in combat support operations. The environment surrounding explosive detection operations does not interfere with the explosive detection capabilities of the PEDD team.

SUSPICIOUS/UNATTENDED PACKAGE RESPONSES

7-14. Once a suspicious/unattended item that may contain explosives has been discovered, the military police or security forces on the scene request a PEDD team. Normally, military police evacuate the area and establish a CP a safe distance away from the threat using the stand-off distance listed in *Appendix I* of this manual. The PEDD team checks the perimeter and secures the area for secondary explosives. EOD personnel should respond to the scene and make contact with the PEDD team to be briefed on the location of the package. The PEDD team moves to within 50 meters of the package and begins an area search up to package. The PEDD team also scans the immediate area for secondary devices, when applicable.

> ***Note:*** At no time should a PEDD team be used to search the package itself. EOD personnel have received special training in handling suspicious/unattended packages and determining if a suspicious/unattended package contains explosives.

7-15. A PEDD handler must not present, touch, open, or physically disturb a package that is suspected of concealing explosives. Once the MWD team has discovered a suspicious item, EOD personnel assume responsibility for the area and take control of the situation in order to render the package safe or dispose of it as necessary.

PEDD Operations in Support of Operation Iraqi Freedom (OIF)

In April of 2003, SGT Raymond Rivera was performing explosives detection missions with his PEDD Kevin in support of OIF near Baghdad, Iraq. SGT Rivera was requested to sweep an area that a support element was in the process of clearing. There was so much debris that a forklift was being used to move large amounts of furniture and building supplies.

The PEDD team would search a portion of the area and then the forklift would come in and haul off the debris searched. After the third load was hauled off, the dog responded to a mattress on the ground with an office table on it. Expecting an explosive devise, SGT Rivera notified the NCOIC of the detail and the area was cleared of all personnel (to a safe distance) and secured. EOD personnel were notified of the suspected explosives and responded to the scene.

SGT Rivera was standing by at the safe area while EOD personnel uncovered two hand grenades that were buried in the sand just below the surface. SGT Rivera said, "Even though it was only two grenades, I was amazed that Kevin found them under all the debris. I'm glad we found them and not our soldiers."

BOMB THREATS

7-16. A PEDD team should be requested for assistance in detecting explosives whenever bomb threats are made against persons or property for which the Army is responsible. The PEDD team's search for explosives begins only after the area or building has been evacuated of all personnel and security has been established. EOD personnel only respond when a suspect item is found; they do not conduct or participate in searches except in support of the USSS. EOD units place a team on standby once the unit is alerted of a bomb threat. PEDD teams coordinate this with the local EOD unit prior to any operation.

7-17. The PEDD team, EOD personnel, and site leadership collaborate to resolve the situation at a nearby CP (see *Appendix I* for stand-off distances). When PEDD teams are located on an installation, their use must be included in the bomb threat planning guidance.

7-18. A systematic search by the PEDD team should include the assistance of a spotter to ensure safety and accuracy. PEDD teams are to be used only to detect the possible presence of explosives.

WARNING:

At no time will the handler touch, open, or otherwise disturb a possible explosive device. IEDs are easily disguised and can be triggered to explode by motion, sound, light, heat, or remote devices.

7-19. Handlers and spotters must not change or otherwise disturb anything in the search. If the lights or any electric device are off or on, do not touch the switch until a thorough search is conducted.

7-20. Ensure that the handler and spotter are provided with available safety equipment including—

- A flack vest (in addition to body armor).
- Flashlights or helmet-mounted lights.
- Gloves.
- Safety goggles.

WARNING:

Always keep track of how much time has elapsed when using a PEDD team at bomb threat incidents. Sufficient time must be maintained to ensure that the team exits the building or area well before the announced or suspected detonation time.

7-21. If the PEDD responds to the presence of explosives, the PEDD handler and spotter note the location and leave the area. EOD personnel then begin measures to render the device safe or remove it. On the request of EOD personnel, the PEDD team may then continue the search until another suspected explosive is found or until the area is cleared by EOD personnel.

7-22. If the search is 30 minutes or longer in duration, explosive training aids should be planted periodically so that the dog can find them. This helps to keep the dog interested in the search. Training aids should only be planted in areas that have already been searched by the PEDD team.

HIGH-RISK PERSONNEL SEARCHES

7-23. Military police provide the task force commander with valuable area security capabilities to protect C2 headquarters, equipment, and services essential for mission success. HRP security is an important task within the area security function. PEDD teams may be employed to enhance explosive detection and protection capabilities.

7-24. The KM or plans NCO determines which PEDD teams will perform the mission. The number of PEDD teams for any mission is determined by the threat, the number of PEDD teams available, any conflicting missions, and other environmental considerations noted in *Chapter 8*.

7-25. PEDD teams may be used as part of a combined arms operation or as an element of a larger security force. When used as part of a combined arms team, the PEDD teams need to be involved will all troop-leading procedures, including rehearsals. All personnel must have a complete understanding of the mission in order to create a safe and secure environment for the HRP. The PEDD teams provide explosive detection and personnel detection capabilities. PEDDs are also a deterrent to hostile forces or terrorists operating within the area of operations.

7-26. Prior to an HRP visit to an area, the PEDD team assists with force protection actions by searching for explosives and IEDs. PEDD also conduct searches of and provide security for outer perimeter areas. During HRP operations, the handler

- Briefs the on-scene commander of the dog's capabilities.
- Communicates with EOD personnel.
- Obtains the HRP's itinerary and determines locations to be searched.
- Requests a spotter or helper to assist.
- Ensures that areas are secure after searches are done.
- Informs the commander of results and continues follow-on missions.

ACCESS CONTROL POINTS

7-27. Using PEDD teams at ACPs can increase an installation's or base camp's force protection and antiterrorism capabilities by providing a visual deterrent to terrorist groups and other criminal elements. The PM notifies the KM of the requirement to perform vehicle searches at ACPs.

7-28. In the event that the PEDD detects the presence of suspected explosives, the military police forces at the ACP secure and evacuate the area (see *Appendix I* for stand-off distances). EOD personnel are then notified to render the item safe or remove it. PEDD handlers do not attempt to remove or otherwise disturb any suspected explosives.

7-29. PEDD team patrol capabilities may be used at ACPs to search for stowaways and personnel trying to gain access to the installation or a controlled area by hiding in delivery vehicles or other large cargo type vehicles. Extreme caution must be used when utilizing the team in this manner. The commander must be aware that once the PEDD is given the command to search for personnel the potential for injury exists.

HEALTH AND WELFARE INSPECTIONS

7-30. PEDD teams conduct health and welfare inspections when requested by the commander and coordinated through the KM, PMO, and S3. The commander contacts the KM directly to schedule a health and welfare inspection. The KM, in conjunction with the installation PM, the CID special agent in charge, and the military police investigations supervisor, briefs all levels of the installation command on the availability and capabilities of the PEDD teams available to them.

7-31. When scheduling the inspection, the commander should have the following information available:

- The requested time of the inspection.
- The areas to be inspected.
- The number of living quarters.
- The number and size of common areas.
- Other pertinent information.

Appendix H provides a guide to assist handlers with health and welfare inspections.

7-32. Once the coordinating instructions have been established, the KM coordinates a time and location to brief the commander and demonstrate the PEDD's capabilities. It is important that the unit requesting the health and welfare inspection practice OPSEC and inform only those personnel who have an immediate need to know. This usually consists of the commander and 1SG or sergeant major. The KM advises the PM and the military police investigations supervisor when the inspection is scheduled and if any additional support is needed.

7-33. A systematic inspection of the living quarters and common areas is conducted. If the PEDD responds to a locked wall locker, the soldier assigned to that wall locker is summoned to open it. Once the wall locker has been opened, unit representatives conduct a detailed search of the wall locker and its contents.

AIRCRAFT AND LUGGAGE SEARCHES

7-34. Before beginning the search, ensure that all personnel have exited the aircraft and that all compartments are opened. The PEDD team should search the aircraft before searching any luggage or cargo. The PEDD team conducts a fresh-air perimeter search from the downwind side when approaching the aircraft. Once a perimeter search has been conducted, the team boards the aircraft at either the front or rear. Once the area has been deemed clear, the handler ensures that only one door is open to limit draft and wind current. A systematic search of all areas within the aircraft is conducted from the top to the bottom.

7-35. If the dog alerts the handler to a possible presence of explosives, military police or security forces evacuate and secure the scene. The PEDD team moves a safe distance away from the suspected explosives but remains available to further assist EOD personnel, if requested. Once EOD personnel clear the area, the PEDD team may resume the search until another suspected explosive is found or until the area is cleared by EOD personnel.

7-36. When conducting inbound baggage inspections, the items should be aligned in a sterile area in rows with at least six feet between them. The handler should conduct an on-leash search, presenting each piece of baggage. When searching cargo, every effort should be made to search the cargo before it is palletized. The cargo should be aligned in a sterile area in rows with at least six feet between them (the same as baggage). The handler should conduct an on-leash search, presenting cargo as required. Once the PEDD team has searched the cargo, it may be palletized while remaining in the sterile area prior to embarkation.

7-37. Conducting debarkation operations is the same as inbound operations. The host unit (to which the cargo belongs or is being delivered) provides a representative (E6 or above) and work detail to assist the PEDD team. When pallets are removed from the aircraft, they are placed in a sterile area. Once in the sterile area, the work detail unloads the pallets and arranges the cargo as required by the PEDD handler.

Chapter 8

Veterinary Support

At most locations, veterinary support is provided by an assigned Army VCO. When there is no assigned VCO, an attending VCO from a nearby military installation is assigned responsibility for providing veterinary support. The Army provides veterinary support for all MWDs as authorized by *DOD Directive 6400.4* and *ARs 40-1, 40-3*, and *40-905*. Emergency, civilian veterinary support is authorized for MWDs when there is not a VCO available and/or care is beyond the capability of the veterinary treatment facility (VTF). The responsible VCO and Army Veterinary Service chain of command are responsible for establishing standard procedures for approval, provision, and payment for emergency, civilian veterinary support.

VETERINARY RESPONSIBILITIES

8-1. The US Army Surgeon General provides professional veterinary support for the entire MWD program through the Army Veterinary Corps. This responsibility includes—

- Medical and surgical care at training facilities, bases, and installations.
- Inspections to ensure that the kennel facilities are sanitary.
- A professional review of plans for new construction and modification of kennels, support buildings, and sites.
- Prescribing an adequate feeding program.
- Instructing handlers and supervisors in all matters related to the health of MWDs.
- Conducting research to improve the MWD program.

8-2. The responsible VCO provides treatment for the diseased or injured MWD at the kennel site or at the installation VTF. The Army Veterinary Corps is responsible for equipping this facility and for providing medical and surgical supplies. The VCO is responsible for the MWD veterinary treatment record (VTR) and related information about examinations, immunizations, and treatment.

8-3. The VCO instructs handlers about dog health, care, feeding, and first aid. This instruction helps the handler to develop a better understanding of the MWD's health needs and improves the handler's ability to care for the MWD. The VCO prescribes an appropriate feeding program based on the MWD's health, the climate, and working conditions.

8-4. The VCO is involved in the review of all plans for new kennel construction or kennel modifications. This ensures that potential health and safety hazards can be corrected before construction begins.

VETERINARY INSPECTIONS

8-5. At least quarterly, the VCO inspects the MWD kennel facility to make sure that proper sanitary standards are being maintained. It is recommended that an animal care specialist conduct a monthly check of the kennels to ensure that no major safety violations are present. The VCO ensures that insect and rodent control efforts are adequate and that the handlers are properly maintaining the general health of the MWD. The quarterly inspection also includes the following elements:

- An examination of the kennel facilities for safety hazards and distractions that may interfere with the rest and relaxation of the MWDs.

- A review of the adequacy of the kennel structure (particularly for environmental conditions) and feeding and watering schedules. The VCO should make recommendations to help prevent disease and injury.
- A review of kennel sanitation and cleaning procedures and MWD food storage procedures to minimize the risk of disease transmission to the MWD.

MANDATORY TRAINING

8-6. The local VCO is required to provide training to the MWD handlers upon initial assignment and at least annually thereafter. More frequent training of handlers is recommended but not required. The Department of Defense Military Working Dog Veterinary Services (DODMWDVS) recommends that handler training be conducted at least quarterly. Topics include the following:

- Procedures on when and how to obtain veterinary support (an SOP is appropriate for this).
- General animal health.
- Grooming.
- Prevention of environmental injury (such as heatstroke or cold injuries).
- Emergency first aid (for bloat [gastric dilatation-volvulus], heat stroke, snake bites, and cold injuries).
- Administration or application of routine medications.
- Proper MWD feeding.
- Feed sanitation (proper storage, preparation, and cleanup).
- MWD chemical, biological, radiological, nuclear, and high-yield explosives protective measures.

8-7. It is best to schedule mandatory training so that seasonal problems are addressed prior to their occurrence and appropriate refresher training is provided prior to a deployment. This training is not meant to turn handlers into animal health technicians, but to give them the tools needed to provide emergency care until they can acquire veterinary support.

WEIGHT CHECKS

8-8. Each month the MWDs body weight must be recorded and sent to the responsible VCO. The KM may communicate this to the VTF. Body weight is also measured and recorded monthly in order to monitor body condition and detect important trends in weight loss or gain that may indicate illness. The KM and/or handler documents data such as food intake, stool consistency, and vomiting in the monthly records.

RECORDS/HEALTH CERTIFICATES

8-9. VTRs are maintained by the local VTF. It is very important that MWD handlers ensure that all health information is annotated in these records.

8-10. An examination is given just before shipping an MWD from one installation to another. The MWDs are examined to detect injury or disease and to support the safe shipment and continued good health and performance of the MWD. A health certificate is issued for the state or country into which the MWD is being shipped for TDY and must accompany the MWD. When the TDY period exceeds 10 days, a health certificate may be required for the MWD's return to his home station. A VCO signs this certificate no more than 7 days prior to reembarkation. If the TDY location is in a different Army VCO's area of operation, the gaining VCO is notified prior to the MWD's departure from the home station. An examination is given as soon as possible, but not more than 10 days after the arrival of an MWD at a new assignment, TDY, or deployed location and within 3 days of the MWD's return to his home station. It is important that the handler who deployed with the MWD is present for the examination so that the VCO can get accurate and complete information on any health issues while the MWD was deployed.

Feeding and Food Control

8-11. MWDs require a diet that is significantly different from that of pet dogs. Their work demands much higher levels of energy and larger quantities of essential nutrients. The standard high-performance diet, contracted and supplied by the GSA for MWDs, meets these nutritional needs and is the only approved feed unless otherwise directed by the attending veterinarian.

8-12. Special diets may be procured and fed to individual MWDs when the VCO determines that other than the standard diet is required. Normally, these diets are available from the VTF, but the MWD unit is responsible for procurement costs.

8-13. Some MWDs have been trained using a food reward schedule that also requires special food. Food reward is used only if an MWD fails to respond to any other type of reward. Procure the special food through normal supply channels or authorized local purchase. The amount of food each MWD should be fed depends on his weight, the amount of activity, and the climate. Local VCOs determine the proper amounts to feed. The responsible VCO must be told when an MWD is trained using food.

8-14. The VCO also prescribes the time of day each MWD is to be fed. This depends on the MWD's duty schedule and the schedule of other kennel activities. MWDs should be allowed 2 hours to eat; leftover food is disposed of within another two hours and feeding pans are cleaned and put away. Never leave uneaten food in the kennel past the authorized 2-4 hour feeding period. If the MWD finishes his meal prior to the end of the feeding period the feed pan may be removed.

Parasite Control

8-15. MWDs are at risk of becoming infested with internal parasites (intestinal worms and heartworms), and external parasites (fleas, ticks, and mange). Some of these parasites may also transmit infectious diseases to the MWD. In order to prevent infestation and disease transmission, all MWDs must be on a routine parasite prevention program supervised by the responsible veterinarian. At a minimum, this must include the administration of a monthly oral (pill or chewable) heartworm preventative and internal parasite preventive and the administration of a topical, veterinary-prescribed external parasite preventative against fleas and ticks.

8-16. When deployed to areas with high parasite risk, additional preventive measures such as prescription flea and tick collars and the oral antibiotic doxycycline will be prescribed by the supporting veterinarian. Doxycycline must be given with food. The handler must ensure that adequate supplies of all preventives are issued by the VTF prior to departing for TDY or deploying. Flea and tick collars will only be used when prescribed by the veterinarian and must be removed when the MWD is not under direct physical control (when off the choke chain and leash) of the handler to prevent accidental ingestion.

Food Storage

8-17. Food should be stored in a sealed container that will not allow contamination or rodents to enter the food (for example, pail, utility 7 gallon, NSN 7240-01-411-0581). Bulk storage of food must be in an area separate from the storage of other equipment, cleaning supplies, or materials that may contaminate the food. Bags should be elevated 4-6 inches above the ground, and neatly stacked with a 4-inch space between the stacks and the walls. The sanitation standards for the storage of MWD rations are identical to those required for the storage of bulk rations for personnel in a troop issue subsistence activity, exchange, or commissary.

Deployability Categories

8-18. The purpose of reporting MWD deployability status is to assist the MWD program managers, PM, commanders, and KMs with the selection of MWD teams for deployment and TDY and to rapidly identify MWDs with limited duty requirements. These guidelines take medical and physical fitness factors into consideration including—

- The anticipated MWD duty stress at deployed locations compared to the home location.
- The availability of veterinary support at the deployed location.

- The climate and environment of the deployed location.
- Any diagnosed or suspected medical conditions that may affect the MWD's performance.
- The MWD's physical conditioning and stamina as observed by the KM and handler.

8-19. It should be noted that some determinations are and will remain subjective; thus the need for prudent judgment by all individuals involved in these assessments remains. This is particularly true regarding the effect of age on an individual MWD. Old age is not a disease and MWDs are not downgraded just due to age; however, with increased age comes increased incidence of illness or disability. These problems may warrant a change in deployability status.

8-20. Deployability status is initially determined after the VCO examines the MWD at the receipt physical or semiannual physical exam. The KM determines training and proficiency status. The VCO and KM should discuss their findings and then categorize each MWD's status and forward the listing through the KM to the PM and MWD program manager. Updates to deployability status lists are performed at least monthly or as needed as changes occur in an individual MWD's status.

Category 1—Unrestricted Deployment (Both Outside the Continental United States and in the Continental United States)

8-21. The MWD is medically fit for any contingency or exercise. There are no limiting or compromising factors. Medical factors may exist but will not limit performance. Special diets, controlled drugs, or special medications can be sent with the MWD (or be carried by the handler) in a quantity sufficient to last the full duration of the anticipated deployment.

Category 2—Limited Deployment (Continental United States or Home Theater of Operation)

8-22. The MWD is medically fit for regions and missions with a minimal requirement for acclimation to heat or physical stress where complete veterinary support is available. The MWD is fit for short duration deployments. There are no significant limiting or compromising factors. Medical problems may exist which slightly limit performances but are controllable. Special diets or medications are required; however they are available in the deployed location or can be sent with the MWD. The reason for limited deployability must be reported.

Category 3—Temporarily Nondeployable

8-23. A medical condition exists that impedes daily duty performance and is under diagnosis, observation, or treatment. The MWD may still be allowed to work at the home station but in a limited capacity.

Category 4—Permanently Nondeployable

8-24. An unresolved medical or physical problem exists that frequently or regularly impedes daily duty performance and an early retirement date cannot be given. The reason for nondeployability must be reported. Medical or physical conditions warrant replacement within one year. The responsible veterinarian and KM should initiate an MWD disposition process at the time an MWD is declared category 4.

Kennel Sanitation Procedures

8-25. Cleanliness is one of the most important factors for the good health of an MWD. The KM must enforce sanitary measures in and around the kennel area. A good standard of sanitation is the result of the cooperation of the handlers, KMs, supervisors, and the VCO.

8-26. The VCO and the PM or military police commander should set the standard of sanitation. This standard must be maintained by each of the handlers and the KM.

8-27. Kennels must be sanitary, in a good state of repair, and thoroughly cleaned every day. Kennels should be disinfected at least once every week using only those disinfecting products approved by the VCO. Kennels should also be disinfected whenever an animal is removed from a kennel so that the kennel will be ready to be occupied by another animal.

8-28. Stool must be removed from the runs as often as necessary as but no less than twice daily to prevent the MWD from becoming dirty. The method of disposing of stool depends on local conditions and the type of sewage system present. If stool must be carried from the area in cans, the cans must be cleaned and disinfected after each use.

8-29. The entire kennel area must be free of refuse and garbage that could attract rats and insects. Mosquito control measures must be used in ditches and swampy areas in the vicinity of the kennels. Disinfectants and disinfectant procedures should be used only with the approval of the VCO.

MEDICATION

8-30. At times, a VCO prescribes special medication (to be given separately or mixed with food) for a sick or injured MWD. The handler must know how to administer these medications in both pill and liquid form.

8-31. When any foreign substance is placed directly into a dog's mouth, his first reflex is to spit it out. The handler must learn how to administer medication properly so that the MWD is forced to swallow. With the VCO's approval, pills or liquids can be mixed with a small amount of canned dog food or other soft food and fed to the MWD. To administer capsules or tablets—

- Place the fingers of your left hand over the MWD's muzzle and insert your left thumb under the lip and between his upper and lower teeth right behind the canine tooth, pressing your left thumb against the roof of his mouth to open it.
- Place the capsule or tablet into the MWD's throat at the extreme rear of the tongue to prevent him from spitting it out. Quickly remove your hand, close the MWD's mouth, and gently massage his throat until he swallows. The entire procedure must be quick and smooth to ease the MWD's apprehension and resentment.

8-32. Liquid medication is best administered with the help of another person. The handler—

- Holds the MWD's upper and lower jaws together with his left hand. The assistant pulls the MWD's lips away from the teeth at one corner of the mouth with his right hand. The MWD's nose is then pointed slightly upward, forming a natural funnel by the lip, and the assistant pours the liquid into this funnel.
- Elevates the MWD's head only slightly above the horizontal when giving liquid medicine. If the MWD's head is raised higher, he has difficulty in swallowing. Give the MWD adequate time for swallowing to prevent the liquid from getting into the trachea, nose, or lungs. Use extreme caution in giving oily liquids.
- Allows the MWD to lower his head and rest before proceeding if any signs of distress, such as coughing or struggling, appear. Do not give oral medications or any liquids if the MWD is unconscious or cannot swallow.

FIRST AID

8-33. Normally, the handler's early recognition of symptoms of illness or injury allows sufficient time to get assistance from a VCO. However, situations may arise when medical help is not immediately available. The seriousness of the incident may require that the handler take emergency actions to protect the life or health of the MWD. The first aid instructions that follow are commonly used to save life, prevent further injury, and reduce suffering from pain. In all emergency situations, notify the VCO as soon as possible so that the MWD can receive professional medical attention.

8-34. First aid kits should always be available in the kennel area and at training sites. The handler should also carry a first aid kit as part of his or her equipment for the MWD on all operational missions. The VCO will determine the contents of first aid kits. When items are used from the first aid kit, they should be replaced with new items immediately.

MUZZLES

8-35. When an MWD has been injured, the first consideration is to calm and immobilize the animal. Pain and distress, however, may cause the MWD to respond to the handler in an unpredictable manner. The MWD may not respond to verbal commands and may attempt to bite the handler and anyone helping the handler. Whether to apply a muzzle (*Figure 8-1)* or not depends on the nature of the emergency. If the MWD is unconscious, shows any difficulty in breathing, or has suspected head injuries, a muzzle should not be used. Otherwise, a muzzle should be used for safety.

Figure 8-1. Muzzle Use

8-36. There are several types of muzzles, but the leather basket muzzle is the best and most comfortable. It allows for free breathing and causes the least alarm and apprehension. The MWD can still inflict a wound while wearing this type of muzzle, so caution is required.

8-37. An improvised muzzle can be made using the MWD's leash and is called a leash muzzle. To apply the leash muzzle, tighten the choke chain on the MWD's neck by pulling the leash tightly with the right hand. Place the left hand, palm up, under the choke chain on the neck. Grasp the leash tightly as it passes through the palm of the left hand, wrap the leash once around the MWD's neck, and bring it up and across the left side of his head. Finally, wrap the leash twice around the muzzle and grab it tightly with the left hand.

CAUTION

The leash muzzle may be used when the leather muzzle is not available or when it would not provide adequate safety. Do not use the leash muzzle when the MWD is overheated, is having difficulty breathing, or is indicating that he may vomit. Do not leave it on for long periods in hot weather.

FRACTURES

8-38. If a fracture occurs, restrain the MWD and encourage it to lie down and remain quiet. If the face and jaw are not injured, put the MWD in a loose muzzle. Notify the veterinarian immediately when a fracture occurs or is suspected. If there is any bleeding, treat the MWD according to *paragraph 8-40,* then proceed with splinting and moving the MWD to the veterinarian if he is not immediately available.

8-39. Prior to moving the MWD, the fracture area needs to be immobilized with a splint. This should be possible for most leg fractures. The leg should be splinted in the same position it is placed by the MWD when it is hurt. Do not try to straighten or reduce the fracture. Splinting an MWD's leg is similar to splinting a soldier's leg and may be done with the same materials: bandages, torn cloth, splints from the medic's bag, or other available material including sticks, boards, and rolled newspaper. After the leg is splinted, or if the fracture cannot be splinted, gently move the MWD onto a litter or improvised litter, such as a piece of plywood, or into a kennel and carry it to a vehicle for transport. It is best to slide the litter under the MWD rather than lifting the MWD onto the stretcher. The MWD may need to be strapped to the litter to allow safe movement. Make sure to support the injured leg or area during movement.

Bleeding Wounds

8-40. Bleeding must be quickly controlled, particularly wounds in the foot or leg that are prone to bleed freely. Applying pressure directly on the wound may control bleeding; use a sterile bandage or clean handkerchief or pinch the edges of the wound with your fingers. As soon as possible, apply a pressure bandage.

Burns

8-41. Most burns occur when an animal comes into contact with hot water, grease, tar, or other scalding liquids. MWDs may also get electrical burns by chewing electrical wires. If an MWD is trapped in a burning building, it may suffer from smoke inhalation in addition to surface burns. The handler can prevent all of these situations through positive control of the MWD. If a burn does occur, applying cold-water soaks or ice packs to the burn for approximately 20 minutes may treat minor burns. The cold helps to reduce the pain. The hair around the burn should be clipped away and the burn should be washed gently with a surgical soap. Blot it dry with sterile or clean soft gauze or cloth. Protect the burned area from rubbing by applying antibiotic ointment and a loose fitting gauze dressing.

Shock

8-42. An animal may go into shock after injuries to internal organs, excessive bleeding, or trauma. Shock can be recognized by the following symptoms:

- A glassy look to the eyes.
- A rapid or weak pulse or rapid shallow breathing.
- A dropping body temperature. The lips and feet may feel cold.
- A paleness of the membranes of the mouth and eyes.
- Slow capillary refill time. To determine capillary refill time, press firmly against the MWD's gums until they turn white. Release the pressure and count the number of seconds until the gums return to their normal color. If it is more than two to three seconds, the MWD may be going into shock. Failure to return to the red-pink color at all indicates that the MWD may be in serious trouble and needs immediate assistance.

8-43. If a handler suspects that an MWD has internal injuries or is going into shock, request help from a VCO immediately. Keep the MWD warm and quiet, and lower his head to prevent possible brain damage. If it is necessary to move the MWD, use a litter. Administration of intravenous fluids by appropriately trained personnel will help to prevent permanent injury from shock.

Artificial Respiration

8-44. There are conditions that may cause an MWD to stop breathing. When this happens, do not panic. Follow these procedures:

- Open the MWD's mouth and check for any obstructions. Extend the MWD's tongue and examine his throat.
- Clear the MWD's mouth of any obstructions then, close his mouth, and hold it gently closed.
- Inhale, and then cover the MWD's nose and mouth with your mouth.
- Exhale gently. Do not blow hard. Carefully force air into the MWD's lungs and watch for his chest to expand. Repeat every five to six seconds or at a rate of 10 to 12 breathes per minute.

SNAKEBITES

8-45. The bite of a poisonous snake can cause serious illness and death if not treated immediately. Keep the MWD quiet and calm, and request veterinary assistance or move the MWD as quickly as possible to a treatment facility. Panic or exertion causes snake venom to move more rapidly through the bloodstream. If possible, kill the snake so it can be shown to the VCO. The handler should be careful not to be bitten by the snake or come into contact with the snake venom. Bites may occur on the face or neck of the MWD. When this happens, remove the choke chain and loosen or remove the collar and muzzle. Swelling occurs rapidly after a snakebite, and this equipment may restrict breathing. Position the MWD's head extended from his body to allow him to get maximum airflow. An ice pack applied to the bite area helps to slow the flow of blood and helps to keep the venom from spreading.

CAUTION

A tourniquet should not be applied. Constricting the blood flow could increase tissue death.

FOREIGN OBJECTS IN THE MOUTH

8-46. An MWD may get a stick or other foreign object stuck in his mouth or throat. The most common foreign object that becomes stuck is a Kong® or ball reward that is too small for the MWD. The MWD may cough, gag, have difficulty swallowing or breathing, paw at the mouth, and drool. If the MWD is having difficulty breathing, cautiously and gently open his mouth. Look for any abnormal object in the throat, under the tongue, between the teeth, in the gums, or in the roof of the mouth. If the MWD is having trouble breathing, removing the object is an emergency. The ball, Kong, or other object can be removed using pliers or clamps. If the object is not causing breathing problems, it is best to leave it in place and get veterinary assistance.

POISONOUS SUBSTANCES

8-47. There are many toxic chemicals an MWD may come in contact with and/or ingest. These include insecticides, herbicides, rodenticides, and antifreeze. The symptoms of poisoning vary. Unless the handler is certain the MWD has eaten a poison, do not treat for poison. Occasionally, MWDs accidentally swallow narcotics or explosives training aids. Exposure to poisons and accidental ingestion of training aides should be prevented by positive control of the MWD by the handler. If the handler knows that the MWD has eaten a poison, narcotic, or explosive, take the following actions:

- Request veterinary assistance immediately.
- Determine the type and quantity of poison, chemical, narcotic, or explosive that has been swallowed. If any part of the substance or container is available, keep it for the VCO to examine.
- If the Army VCO is not available, immediately make contact with a civilian veterinary to determine if vomiting is appropriate. Causing an animal to vomit is not recommended in all types of poisoning as it may cause more harm to the animal. When vomiting is appropriate, the animal may be induced to vomit by giving him one-quarter to one-half of a cup of hydrogen peroxide.
- Keep the MWD quiet and warm until the VCO arrives.

8-48. For future reference, keep a copy of the material safety data sheet for any training aide that may be ingested available. MWD first aid should be according to the material safety data sheet for people.

OVERHEATING

8-49. Overheating results when an MWD is unable to eliminate body heat rapidly enough. This condition requires immediate action by the handler to save the MWD's life. During hot, humid weather, an MWD may easily become overheated during training, during operations, or while being transported. A body temperature of 105° Fahrenheit or more, combined with poor response to commands, weakness, unsteady movement, vomiting, difficult or labored breathing, convulsions, and collapse are all symptoms of overheating. It is helpful to take the MWD's temperature in order to monitor the severity of overheating and progress in cooling, but temperature alone does not determine the need for or method of treatment.

8-50. For mild to moderate heat injury, the MWD's temperature is generally lower than 104° Fahrenheit. The MWD may have mild symptoms, to include uncontrolled panting, a rapid but somewhat weak pulse, mild to moderate difficulty breathing (but not cyanotic [blue]), and exercise intolerance or loss of stamina. This is equivalent to heat exhaustion in people.

8-51. When the symptoms of overheating occur, loosen the leash, collar, and muzzle as much as possible without loosing control of the MWD. This allows for more efficient panting and cooling. Follow these guidelines at the onset of symptoms:

- Carry the MWD to the nearest shade and fan him if no wind is present.
- Try to quickly lower the MWD's body temperature by running and sponging cool water over the MWD's head, body, and legs. If a body of water is available, allow the MWD to stand in the water but make sure the MWD's head remains above water. Running water works better than standing water.
- Place the MWD in an air-conditioned kennel, vehicle, or shelter.
- Apply ice packs to the MWD's groin, armpits, and head to increase speed of cooling.
- Dampen, but don't soak, the MWD's feet, ears, and abdomen with rubbing alcohol. Allow it to evaporate and then reapply more.
- Do not let the MWD over drink. He should not consume more than one cup of water until he is calm and his temperature returns to normal.
- Monitor the MWD's temperature every 5-10 minutes, and stop cooling efforts when it reaches 100-103° Fahrenheit. Keep monitoring his temperature for an additional 60 minutes.

8-52. If the MWD's temperature remains normal (about 100-103° Fahrenheit), treatment is complete; however the MWD must be examined by a VCO and a heat injury must be reported in his VTR. If his temperature climbs again, restart cooling efforts.

8-53. If the MWD's temperature drops below 100° Fahrenheit, begin warming by drying and wrapping him with sheets and blankets. Remove the heat when his temperature goes above 101° Fahrenheit.

8-54. For severe heat injuries, the MWD's temperature is generally over 104° Fahrenheit combined with poor response to commands, weakness, unsteady movement, vomiting, difficult or labored breathing, uncontrolled panting, convulsions, collapse, a dull or depressed attitude and behavior, a rapid but somewhat weak pulse, exercise intolerance or loss of stamina, and cyanosis (blueness) or dark red mucous membranes. This is equivalent to heatstroke in people. If this occurs take the following actions:

- Initiate all cooling and monitoring efforts as noted in *paragraph 8-51*.
- Administer 1 liter of intravenous fluids to the MWD as quickly as possible if appropriately trained personnel are available.
- Continue cooling and monitoring until a veterinarian can evaluate the MWD.

BLOAT (GASTRIC DILATION-VOLVULUS)

8-55. Bloat is an acute stomach enlargement and twisting that may be due to gas, food, or water. It may occur if the MWD is fed immediately before or after hard exercise or when he is returned to his kennel after work or exercise and allowed to drink too much water.

8-56. An enlargement may be seen just behind the ribs, primarily on the left side. The MWD may also be restless and show signs of pain in the abdominal region. He may attempt to vomit or have a bowel movement. His breathing may be difficult or labored due to pressure from the enlarged stomach.

8-57. Handlers should notify the VCO and stop all watering and feeding. Walking may enable the MWD to relieve himself through bowel movements or passing gas. Most cases require extensive treatment by the VCO.

8-58. To prevent bloat, MWDs should not be fed within a two-hour period before or after hard work or vigorous exercise. It is also safer to provide multiple small meals each day, feeding twice daily, rather than one large meal. In hot weather, give small amounts of water during training or working MWDs to prevent excessive thirst. For the first hour after working or training, only three inches of water in the bucket should be available. After this cooling-off period, more water may be given.

FIRST AID KIT

8-59. The responsible VCO determines the contents of MWD first aid kits. The VCO should consider the skill of the handlers, access to veterinary support, and emergency situations that may occur. The DODMWDVS does not have a standard first aid kit supply list because needs vary. The local VCO and KM are responsible for developing and putting together an appropriate kit and training the handlers how to use it. The kit should contain supplies for cleaning and dressing minor wounds, cleaning and protecting an eye injury, bandaging or splinting minor breaks, and removing mild superficial foreign bodies (splinters or glass). It should also contain peroxide, antibacterial solutions, and a small supply of any prescription medications needed for individual MWDs.

ENVIRONMENTAL EFFECTS ON THE MILITARY WORKING DOG

8-60. Just as the environment affects soldiers performing their duties, the MWD is affected by his surroundings. It is important for planners and leaders to become aware of the many different conditions that can affect the performance of MWDs. Not all MWDs are affected in the same manner. Acclimation times will vary between conditions and MWDs. Planning considerations should always include the VCO when the MWD will be used or trained in environments that he is not accustomed to.

WIND

8-61. Always consider the direction and velocity of the wind when using an MWD for any detection operations. This consideration holds true even indoors. There may be a slight breeze that will move a scent away from the location of its source. Fans, air conditioning units, and objects in motion can create winds that will affect the performance of a detector dog.

8-62. In addition to these operational concerns, wind may create medical issues (such as blowing sand) that are negative. Wind also creates increased cooling, which is positive in hot weather but negative in cold weather.

INCLEMENT WEATHER

8-63. Precipitation can degrade the MWD's performance of detection operations, however, do not assume that an MWD will not be able to perform satisfactorily. The amount of moisture accumulated and the amount of time elapsed since the scent was left are variables. MWDs have been proven to be effective even in inclement weather. MWDs that become wet and are returned to their kennels or crates while wet are at increased risk of skin injury and infection.

HEAT

8-64. When operating in locations of extreme heat, the MWD should be allowed sufficient time to acclimate. Consult with the VCO at the location to determine how long an MWD should be allowed to acclimate. Care should be given to ensure that the MWD does not sustain injuries to the pads of his paws due to excessive walking on concrete or pavement in extreme heat conditions. Always ensure that the MWD has a fresh and constant source of drinking water.

COLD

8-65. Extreme cold temperatures should not deter MWD operations. Special care should be taken to enable an MWD to perform in cold locations. The accumulation of moisture on the MWD's paw pads can freeze and injure him. Using paw pad protectors to aid in the prevention of injury due to cold weather may be helpful, but the MWD must be trained to wear them. The MWD should be allowed to rest and warm up on the same cycle as the handler.

TERRAIN

8-66. The safety of the handler and MWD must be given the highest consideration when operating in hazardous terrain. Care should be taken when employment will take place in rocky or mountainous areas. Unfamiliar locations can pose threats in the form of caves, holes, loose footing, and other hazards. Whenever possible, conduct a reconnaissance of the area before employing an MWD.

DISTRACTIONS

8-67. There are numerous distractions that can affect MWD operations. These are conditions that can cause the MWD to focus his attention away from the desired task at hand. Planning and consideration should be given to eliminate or reduce distractions that hinder operations. Distractions are natural and man-made. Some examples of distracters are:

- Other dogs (both MWD and indigenous).
- Other animals.
- Human presence and activities.
- Noise.
- Smells.

Appendix A

Metric Conversion Table

This appendix complies with current Army directives, which state that the metric system will be incorporated into all new publications. *Table A-1* is a conversion chart.

Table A-1. Metric Conversion Chart

US Units	***Multiplied By***	***Equals Metric Units***
	Length	
Feet	0.30480	Meters
Inches	2.54000	Centimeters
Inches	0.02540	Meters
Inches	25.40010	Millimeters
Miles (nautical)	1.85320	Kilometers
Miles (statute)	1.60930	Kilometers
	Area	
Square inches	6.45160	Square centimeters
Square feet	0.09290	Square meters
	Volume	
Cubic feet	0.02830	Cubic meters
Fluid ounces	29.57300	Milliliters
	Temperature	
Degrees Fahrenheit	Subtract 32, multiply by 5/9	Degrees Celsius
	Weight	
Ounces	28.34900	Grams
Pounds	0.45359	Kilograms
Metric Units	***Multiplied By***	***Equals US Units***
	Length	
Centimeters	0.39370	Inches
Kilometers	0.53960	Miles (nautical)
Kilometers	0.62137	Miles (statute)
Meters	3.28080	Feet
Meters	39.37000	Inches
Millimeters	0.03937	Inches
	Area	
Square centimeters	0.15500	Square inches
Square meters	10.76400	Square feet
	Volume	
Cubic meters	35.31440	Cubic feet
Milliliters	0.03380	Fluid ounces

Table A-1. Metric Conversion Chart (Continued)

Metric Units	*Multiplied By*	*Equals US Units*
	Temperature	
Degrees Celsius	Multiply by 9/5, add 32	Degrees Fahrenheit
	Weight	
Grams	0.03527	Ounces
Kilograms	2.20460	Pounds

Appendix B

Kennel Construction

This appendix is intended to be a starting point in the design process with the identification of requirements and spaces. It should serve as a supplement to the installation design guide and not as a replacement. This appendix provides commanders and planners with guidelines and specifications to consider for the construction and equipment of kennel facilities. The floor plan (*Figure B-1*) is an example of how an MWD facility works, how areas relate to each other, and what is required for each space. It is not intended to serve as a rigid model for duplication.

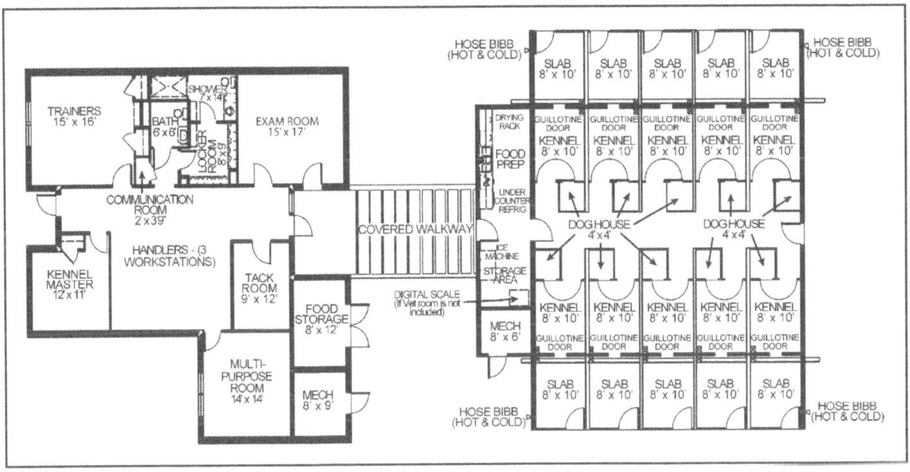

Figure B-1. Example of an MWD Kennel Floor Plan

TEMPORARY KENNELS

B-1. There are two major types of kennels, temporary (*Figure B-2*, page B-2) and permanent. According to *AR 190-12*, temporary kennels are to be used during deployments for no longer than one year, to allow sufficient time for permanent kennels to be built. Temporary kennels may be constructed out of any available resources. The design is not standard and varies from site to site depending on the location, mission, and duration of operations. Using a concept similar to constructing fighting positions, the temporary kennels will be improved continuously as needed or until a permanent facility is completed.

B-2. Using tents and shipping crates may be a hasty method of setting up temporary kennels. Although this method is not optimal for sustained operations, it may be useful until other accommodations can be coordinated. Erect the tents at a location that allows the MWDs a safe place to rest away from other troop activities after operations. Whenever possible, use terrain or buildings to place a barrier between troop activity and the temporary kennel. Never place the temporary kennel site near motor pools, dining facilities, or high noise areas.

Figure B-2. Temporary Kennel

B-3. In some areas of operation, it may be possible to obtain the use of a building or other structures for temporary kennels. In these cases, leaders must ensure that the building is well ventilated, safe, and structurally sound. Ensure that there are no hazards (such as chemicals, bare wires, holes, or debris) that can cause injury to the MWDs or handlers.

B-4. Engineers are often used to assist in the improvement of temporary kennels. In addition, fences should be erected and dog runs created.

***Note*:** From the start of operations, sanitation measures must be employed in all temporary kennel configurations.

PERMANENT KENNELS

B-5. The following information and figures give specific details for the construction of a permanent kennel facility and adjacent areas. These guidelines may be altered to allow for special needs or uses unique to each location.

EXTERIOR AREAS

B-6. An MWD kennel complex (*Figure B-3*) consists of kennels, a support building, an obedience course, an exercise area, an MWD break area, exterior storage, parking areas, drives, and walks. Often, as a result of the location, utility services are limited at the site. Construction should provide the following utilities:

- Water (potable with backflow prevention).
- A sanitary sewer.
- Electric.
- Telephone.
- Network and communication service to the site.

Other systems should be incorporated according to local procedures.

B-7. The entire complex should be enclosed, at the minimum, with a heavy-duty, 8-foot-high chain-link fence with three strands of straight wire (no barbed) at the top to prevent an MWD from climbing or jumping out. A 10-foot (minimum) vehicle gate that can be padlocked is to be installed to allow for food deliveries to the kennel and other access requirements. A personnel entry gate that is visible from the KM's office is also required. The personnel gate should have a cipher lock and a notification buzzer that sounds in the KM's office, the obedience course, and the kennel. The buzzer in the kennel should have the capability to be turned on when personnel are in the area and turned off to allow the MWDs to rest when personnel are absent. All gates should be self-closing and self-latching.

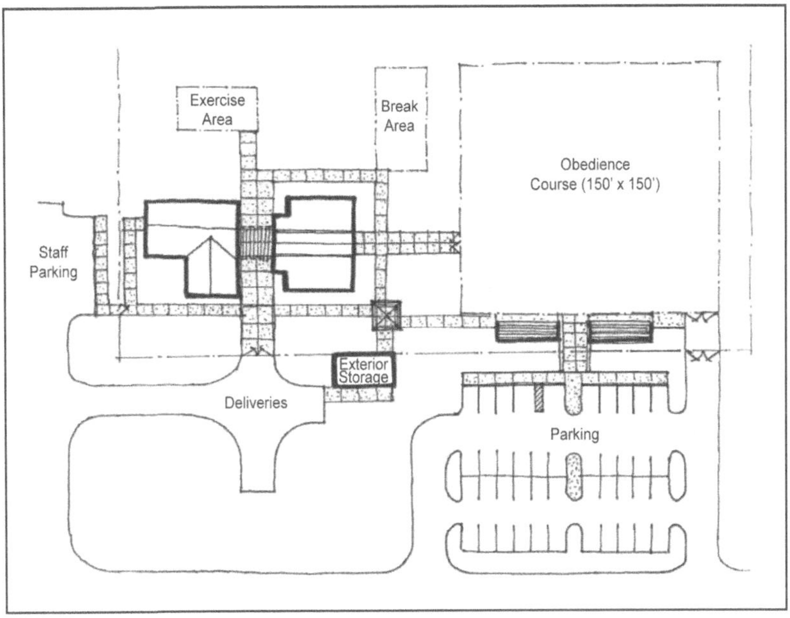

Figure B-3. MWD Kennel Complex

B-8. Security lighting for the area should be according to local policy. At a minimum, the lighting should be on from dusk to dawn and placed on all corners of the fencing area. Security lights should be placed in areas that the other lights do not cover.

B-9. Warning signs will be posted on the exterior fencing and buildings of the MWD kennel and exercise area. Signs should contain the following words: "DANGER—OFF LIMITS–MILITARY DOG AREA." Personnel approaching the kennel area should be able to see and read the warning signs under normal daylight conditions from a distance of 50 meters.

B-10. Locate a trash dumpster for the complex outside of the perimeter fence. Screen the dumpster and provide low-maintenance landscaping for the complex according to local policy.

Exterior Storage, Parking Areas, Drives, and Walks

B-11. An enclosed, 400-square-foot exterior storage building is required to store maintenance equipment, portable kennels, and obstacles. Interior/exterior lighting, power, and a water faucet should be included.

B-12. Staff parking should be a paved parking lot large enough for eight POVs, three GOVs, two visitor vehicles, and one handicap parking space. Pave access drives to accommodate parking, exterior storage, food deliveries, emergency vehicles, and the transportation of MWDs.

B-13. Provide paved walks to all areas. When designing, give careful consideration to creating a one-way MWD traffic system. To prevent confrontations between the MWDs, avoid situations that allow MWDs to meet head on.

Obedience Course (150 Feet by 150 Feet)

B-14. The obedience course plays an important role in maintaining the MWD's agility and stamina as well as in reinforcing obedience and proficiency training. The course should be grassed and free of hazards (trees, large rocks, holes, and burrs) that may be harmful to MWDs and handlers. The site should be graded for drainage but minimally sloped to provide a level field for training. The area should be enclosed with an 8-foot chain-link fence. Gates should be self-closing and self-latching with a minimum width of 5 feet to allow for lawn maintenance equipment. The area should be well lighted to eliminate shadows. Light poles should be located on the exterior of the fence. On and off switches should be manual. Weatherproof outlets should be provided at each light to provide power at the site. Speaker units connected to the chosen public address system should also be provided. Water should be at the course site with water faucets installed to provide water for MWDs and lawn watering. An irrigation system is desirable for long-term lawn maintenance.

B-15. Obstacles for the course (*Figure B-4*) include barrels, tunnels, steps, jumps, framed windows, and dog walks. Each obstacle is 15 to 20 feet from the previous obstacle; the course is run in a sequence. Obstacles should be constructed to the following specifications:

- Barrel 1 is 35 inches long with a 23-inch opening.
- Barrel 2 is 70 inches long with a 23-inch opening.
- Barrel 3 is 105 inches long with a 23-inch opening.
- The tunnel is 146 inches long with a 19-inch opening.
- The steps are constructed out of wood and are 94 inches high with 5 steps on each side. Each step is 43 inches wide by 16 inches high by 24 inches long. The top platform is 48 inches long by 43 inches wide.
- Jumps 1, 2, and 3 are constructed of wood and are 36 inches high at the maximum raised level. There are 6 removable boards that are 51 inches long by 6 inches high and 1 inch thick.
- The framed window is constructed of wood and is 8 feet tall. The maximum height for the MWD to negotiate is 36 inches. The window opening is 48 inches wide. Removable boards are 5 1/7 inches wide by 51 inches long and 1 inch thick.
- The A-frame is constructed of wood. Its two sides are 8 feet long by 4 feet wide by 1 inch thick with 1-inch by 2-inch boards every 6 inches to secure footing. It has a maximum raised height of 6 feet with a base spread of 10 feet.
- The dog walk is constructed of wood and is 26 inches high by 224 inches long and 11 inches wide with 1-inch by 2-inch boards every 4 inches to secure footing. Each end has a 45° incline ramp that is 11 inches wide by 49 inches long (included in the 224 overall length measurement).

Note: See *DA Pam 190-12* for more complete construction guidelines including the bracing necessary for the various obstacles.

Military Working Dog Break Area (10 Feet by 20 Feet)

B-16. An MWD break area should be located near the kennels. A break area allows the handler to release the MWD immediately after exiting the kennels or before entering the kennels so that the MWD can relieve himself. The area should be enclosed with an 8-foot chain-link fence. Gates should be self-closing and self-latching and a minimum of 5 feet wide to allow for lawn maintenance equipment. The ground can be a sandy area for cleaning ease. A water faucet is required to maintain the area.

Exercise Area (20 Feet by 40 Feet)

B-17. The exercise area is a space where the MWD can be released without the handler being present. The exercise area should not be a shared area with the obedience course, as it would conflict with training objectives. The exercise area should be visually separated from the obedience course to prevent MWD from being distracted during training. The exercise area should be grassed, hazard free, and graded slightly for drainage purposes.

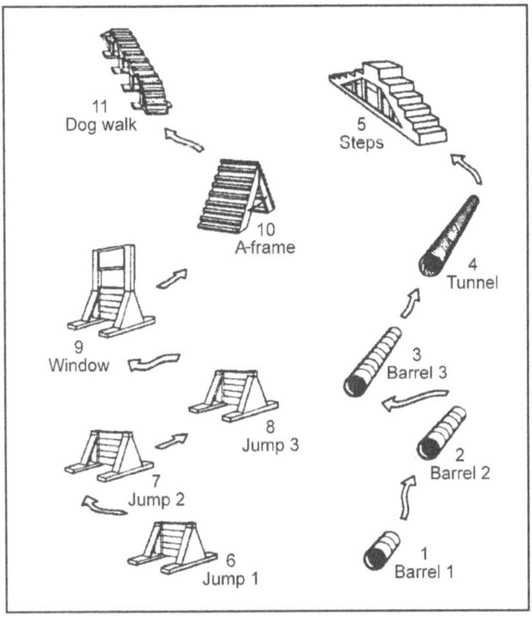

Figure B-4. Obstacle Course Layout

INTERIOR AREAS

B-18. Interior areas have heating, ventilation, and exhaust systems; general-purpose lighting; and power outlets in all spaces. Provide phone, fax, and local area network (LAN) connections to support each area and hot and cold water to all latrines, sinks, and showers. Construct the facility to meet all fire protection and life safety feature requirements.

Administration Areas

B-19. Since MWDs are likely to visit the administration area (*Figure B-5*, page B-6) at some point, finishes should be of durable material, require minimal maintenance, and be easily repaired or replaced in case of damage. (For example, flooring should be covered by carpet.)

Kennel Master's Office (132 Square Feet)

B-20. The KM's office serves as the nerve center for the complex. Ideally, the office should be located at the front of the administrative area with exterior windows that view the entry gate to the complex. It includes a lockable closet for supply storage. Furnishings include a desk, a chair, two visitor chairs, a four-drawer file cabinet, and a bookcase. Equipment includes a computer, a printer, a fax/copier, and a public address system or central intercom console.

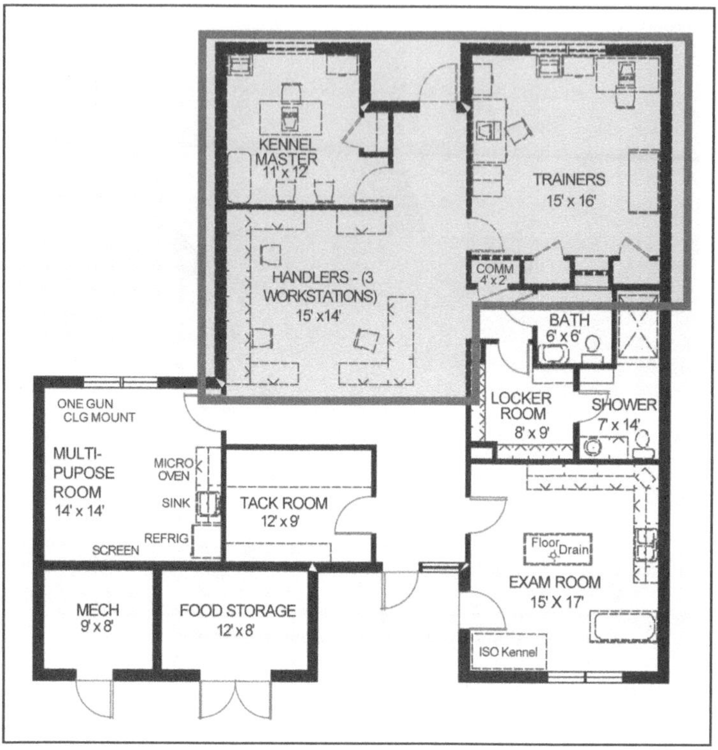

Figure B-5. Kennel Administrative Areas

Trainer's and Plans Noncommissioned Officer's Office (240 Square Feet)

B-21. The trainer office should provide workspace for two people. Lockable closets are required to store training aides for each trainer. Furnishings include a desk, a chair, and a four-drawer file cabinet for each trainer and a common bookcase. Equipment includes a computer, a printer, a copier, and a 4- by 8-foot dry erase board.

Handlers' Area (210 Square Feet)

B-22. The area for the handlers is an open area for general office functions. It includes—

- Three common workstations, depending on the size of the kennels, for handlers to update records and complete daily reports.
- Space for document storage.
- Furnishings (three workstations, file cabinets [one four-drawer vertical file per workstation or the equivalent], and a distribution counter.
- Equipment (computers and a printer).

As a general rule, one workstation should be available for every three handlers assigned to an MWD section.

Controlled-Substance Storage Room

B-23. This room is inside the administrative area and allows sufficient room for the storage safe in which controlled-substance training aids are stored. The doors are secured using approved locking devices. Lightweight doors are replaced with solid metal or wood doors or covered with a 9- to 12-gauge security screen or 16-gauge sheet steel. They are fastened with smooth-headed bolts and nuts and peened in place. All windows providing access to a storage room that is not staffed 24 hours a day are protected by a 9- to 12-gauge security screen or 3/8-inch or larger diameter steel bars spaced no more than 6 inches apart. The frames holding the screen or bars must be fastened to the window frame with smooth-headed bolts.

Special-Use Areas

B-24. Special-use areas (*Figure B-6,* page B-8) are the transition spaces between the administration area and the kennels. They are important in the day-to-day operations of the MWD facilities. They are not normally occupied on a daily basis. As such, these areas provide a secondary function of serving as a noise buffer between the kennels and the administrative areas. The special-use areas provide a workspace for the VCO and accommodate the unique storage requirements for the kennels.

Veterinary Treatment Room

B-25. A veterinary treatment room is used to perform health care and first aid for MWDs since many locations do not have veterinary treatment areas in close proximity to the kennels. In order to maintain sanitary conditions, treatment rooms require a higher degree of cleanliness and durability than the rest of the support building. Basic finish requirements are—

- Seamless floors with an integral base.
- Washable walls.
- Painted gypsum board ceilings.
- Cabinets faced with plastic laminate.
- Solid-surface countertops.

B-26. Interior partitions should provide a sound transmission class rating of 50 to 55. Floor drains are centered in the room to aid in cleaning, and a dedicated floor drain is located under the isolation kennel. Wall and base cabinets with drawers are installed, including stainless steel dual washbasins with hot and cold water. One section of wall cabinets should be lockable so that prescription medicines for MWDs may be stored in the room. In addition to the normal electrical outlets required by code, a 220-volt outlet is required to support other portable equipment that may be used. Equipment for the room includes the following:

- **An MWD isolation kennel.** The entire structure should be made of uniform, nonglare type 304 stainless steel. The frame, cross members, and vertical spokes should be welded together at every intersection. Latches and hinges are made of 12-gauge stainless steel. The floor of the pen is a raised, removable grate. The floor grate is a heavy-gauge stainless steel, plastic-coated section installed 1 3/4 inches off the floor. A dedicated floor drain is required for this kennel.
- **Standard table tub.** The tub is made of type 304 stainless steel with a minimum depth of 15 inches. The tub slopes to the drain end. The design includes a removable, bathing rack made of plastic-coated stainless steel that fits into the bottom of the tub to give MWDs better footing during treatments. Faucets include a flexible, stainless steel hose that will extend the length of the tub and spray powerfully to reach all areas of the tub.
- **Exam room light.** A ceiling-mounted exam room light with an arm extension and swivel capacity of 360° is recommended.
- **Walk-on platform scale.** The scale is a type 304 stainless steel platform with a wall- or post-mounted light-emitting diode (LED) display. The scale is able to perform calibration and weigh animals to the nearest 0.1 of a pound or kilogram.
- **Stationary exam table.** The table top is type 304 stainless steel die-formed from one piece of 20-gauge steel. It should have raised edges to prevent fluid runoff and be permanently attached

to a heavy base. There should be a minimum 35-inch work height from the floor to the top of the table. The heavy base should have adjustable leveling screws.

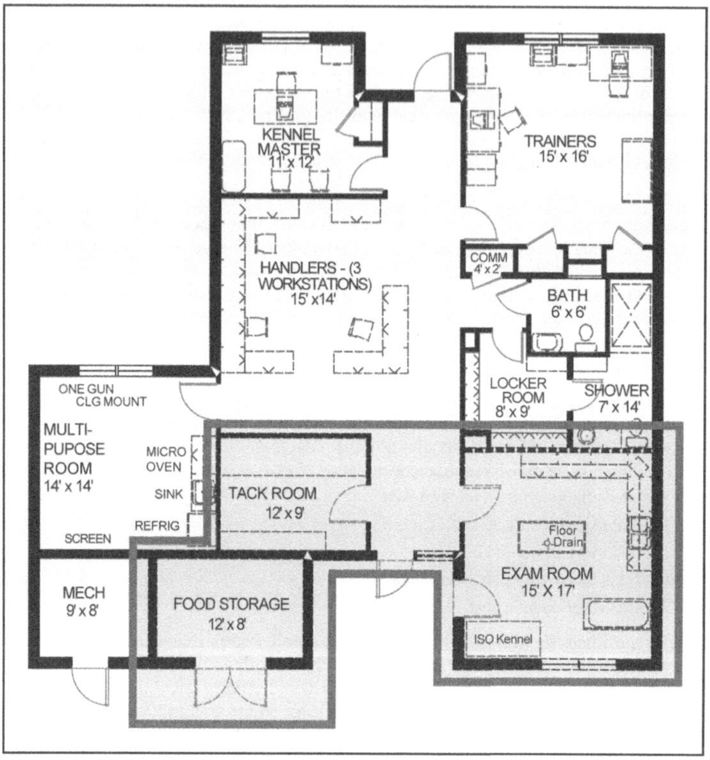

Figure B-6. Kennel Special-Use Areas

Tack Room (108 Square Feet)

B-27. The tack room is for the storage of extra MWD equipment, such as bite suits and portable kennels. The tack room is located near the entrance to the kennels but not the main building entrance. Requirements include 12 linear feet of shelves that are 18-inches wide and 36 inches deep and allow for 36 inches above each shelf; a 24-inch space with hooks to hang items on; and 6 linear feet of shelves that are 18 inches wide and 12 inches deep and allow for 18 inches above each shelf.

Food Storage Room (96 Square Feet)

B-28. The food storage room is normally for the bulk storage of a 30-day supply of food for the MWDs. The room is adjacent to the kennels with an exterior entry. The entry is a pair of 36-inch doors to allow for palletized food delivery and storage. Doors require weather stripping and thresholds to aid in insect and rodent control. Temperature and humidity control equivalent to the office area is required to control food spoilage. A minimum separation of 2 inches is required between the walls and the stored food to allow for air circulation. Shelving should be installed to meet local delivery operations.

SUPPORT AREAS

B-29. The support areas (*Figure B-7*) provide space for the common-use areas of the building. Areas for the multipurpose room, latrines, communications, and mechanical necessities form the core of the support areas. Support areas should be centrally located and easily accessible for building occupants or maintenance personnel in the case of the mechanical room.

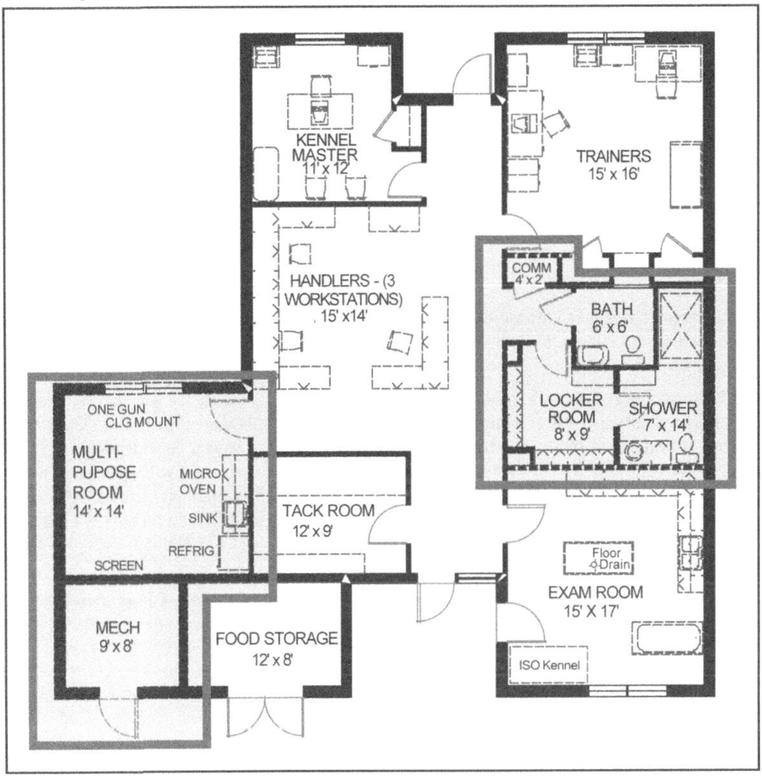

Figure B-7. Kennel Support Areas

Latrine, Shower, and Lockers (206 Square Feet)

B-30. A unisex area is divided into three rooms for the latrine, shower, and lockers. One room is a handicapped accessible latrine with a water closet and lavatory for general use. The locker room includes 10 full-height metal lockers and a bench for MWD handlers' use. The shower room includes a shower, bench, water closet, and lavatory. Do not neglect to include other accessories such as mirrors, soap dispensers, paper towel dispensers, disposal units, grab bars, and hooks.

Heating, Ventilation, Air Conditioning, Mechanical, Electrical, and Communications Systems

B-31. Heating, ventilation, and air conditioning (HVAC) is a process to add hot or cold air in a facility. It also introduces fresh air in to the facility to maintain indoor air quality and eliminate sick building syndrome. Design HVAC systems to optimize energy use and meet the design criteria for the area in which you are located. Place the hot water heater in the mechanical room and with a capacity that will meet the demands of the latrine and shower; multipurpose room; and veterinary treatment room. Locate electrical panels in this area as well. Although a recommended size is shown on the plans, the room should be big enough to support the equipment required for the location.

B-32. While the number of instruments and computers is not large, the communications closet is vital to the success of the organization. Provide a minimum of one prewired telephone and LAN outlet for each workstation, one in each common area, and one in the kennel food preparation area. Include telephone and LAN outlets for fax machines and printers. Conceal conduit in the walls.

Multipurpose Room (196 Square Feet)

B-33. The multipurpose room serves as a combination conference room, classroom, and break room. Food preparation requirements are a light, a countertop, and base and upper wall cabinets for storage. Lighting should have a variable-level control for use as a conference room and/or classroom. Furnishings include either a conference table for fifteen or three five-person tables; chairs; a bulletin board; and a 4- by 8-foot dry erase board. Equipment includes a video projection system, a projector screen, a sink, a microwave oven, and a refrigerator.

KENNEL AREA

B-34. The kennel area provides the daily living environment for the MWDs, giving each MWD a private place for eating and resting. Kennel areas should be built to accommodate large-breed dogs and designed as modular structures to allow for the future expansion of runs.

B-35. There are three types of kennel areas; indoor kennels, outdoor kennels, and indoor/outdoor kennels. The preferred standard is the combination indoor/outdoor kennel; however, there are factors that must be taken into consideration when selecting the kennel type. Cost is always a consideration, but climate also influences the selection. In cold weather climates, the benefit-to-cost ratio will drive the selection to an indoor kennel while the opposite is true for hot weather climates. The more temperate the climate the more suited the indoor/outdoor kennel is for use. Each type of kennel includes the same space requirements. A kennel includes kennel runs with doghouses, a food preparation area, a storage area, and a mechanical room.

B-36. The kennel should be separated from the administrative building by a minimum of 20 feet. This provides separation between different functions and aids in noise control. As a result of the constant interaction between the two facilities, a covered walkway is required to connect the facilities and provide protection from the elements. Care should be taken in the layout of the kennel in relation to the administrative building; a one-way system should be maintained in the movement of MWDs. The objective is to avoid situations where MWDs meet face to face and to prevent confrontations. Consider installing mirrors at corners, intersections, and blind spots to alert handlers of potential wrong-way traffic.

B-37. All entries and exits from the kennel area are required to be self-closing and self-latching. Doors exiting the kennel runs should open inward to aid in preventing MWD escape. Proper ventilation is important in the kennels to prevent the spread of diseases and to control odors. It is recommended that the ventilation standard be to cycle out the room air 10 to 15 times per hour. Kennel temperature should range from 45° to 85° Fahrenheit with humidity in the range of 40 percent to 70 percent. Temperature should be maintained within 10° of the exterior temperature. MWDs work more effectively and are more alert when the kennel temperature is close to the temperature of their working environment.

B-38. Potable water is required at the kennels. Hot and cold water lines are installed at ceiling level down the length of the 5-foot wide center corridor connecting to hose reels at each end of the corridor. A high-pressure washer is required to assist in sanitizing and cleaning kennel runs.

B-39. Since the kennel is basically a wet environment, all electrical receptacles should be provided with ground fault circuit interrupters and all-weather covers. Receptacles mounted at 48 inches above the finished floor are required at each end of the central corridor.

B-40. Each kennel run should have a waterproof light fixture that has an individual switch at the corridor entrance. Other areas require general-purpose lighting according to industry standards. All drain lines in the kennel should be a minimum of 6 inches in diameter and be designed to sustain flow velocities that will maintain a self-cleansing action. Due to the nature of the waste, cleanouts should be numerous and easily accessible. Floor drains should be included in the central corridor to aid in the cleaning of kennels.

Indoor/Outdoor Kennel

B-41. A combination indoor/outdoor kennel (*Figure B-8*, page B-12) simply implies that there is an interior run and an exterior run linked with each MWD living area. A guillotine type door will connect the interior run and the exterior run. Other door types may be used, but the idea is to permit isolation of the MWD during cleaning operations. Ideally, operation of the door is from the central corridor and from the outside of the exterior run. At 6 feet above the finished floor, use translucent wall panels in the wall separating the runs to provide natural light inside the kennel.

B-42. The roof should extend over the exterior runs and include gutters and/or downspouts to prevent excessive water in the runs. Exterior framing should be enclosed with a soffit to prevent birds from roosting. Exterior water faucets (with freeze protection where required) with both hot and cold water are required at each corner of the building. These are necessary for cleaning exterior runs. Floor drains (*Figure B-9*, page B-12) should be located in the corner of each run adjacent to the common wall separating exterior and interior runs. Drains should tie-in to a common waste line. Drain covers should be flush with the floor. Kennel floors should slope (at least 1/4- inch per foot) toward the drains to allow for quick water drainage and drying.

Indoor or Outdoor Kennel

B-43. The plan for the indoor or outdoor kennel (*Figure B-10*, page B-13) is the same for either style. The obvious difference is the wall that encloses the kennels. For indoor kennels, the exterior walls along the runs should have windows beginning at 6 feet above the finished floor to allow for natural light. Windows should be operable and hinged at the sill to tilt inward to prevent the escape of MWDs. For outdoor kennels, the exterior wall is simply an 8-foot chain-link fence. Water faucets with hot and cold water are required at each corner of the building. These are required to clean the gutter and drain system (*Figure B-11*, page B-13) for the kennel. A gutter and drainage system is provided at the back of the kennel run for either type of kennel for sluicing wastewaters from cleaning operations. Kennel run floors should slope to the gutter and the gutter should slope to the drain. Gutters should be sufficient to allow for easy cleaning. A minimum of 3 feet should be allowed between the end of the kennel and the wall or fence.

Kennel Run

B-44. Runs should be arranged so that openings are staggered in order to avoid MWDs facing each other across the corridor. Full-height partitions are required between runs for all kennel types. The first 6 feet of wall above the floor should be sealed concrete. For interior runs, steel clad acoustical panels should be considered from the top of the concrete partition to the ceiling. Steel clad acoustical ceiling panels should also be considered above the kennels.

B-45. Heavy-gauge chain-link fence is used for exterior runs and is also an option for interior runs. Because of the noise generated in the enclosed area of an interior run, it is important that noise-reducing materials be introduced to prevent stress to the MWD and hearing loss to personnel.

B-46. As a minimum, end walls should be full-height, galvanized chain-link fencing. Take care to prevent sharp edges from bolts, hinges, and tie wires on the interior of the runs where MWDs may be cut or harmed. Self-latching entry gates should swing 180° and be able to cover the opening of the MWD house. This allows the MWD to be penned in the house during run-cleaning operations. A stainless steel water bucket is required for each kennel, including a holder that prevents the MWD from overturning the bucket.

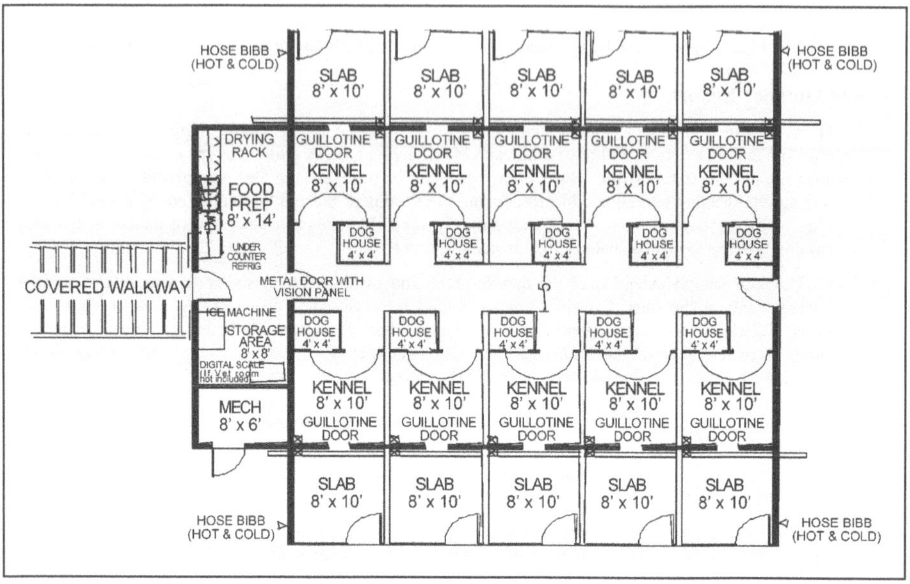

Figure B-8. Combination Indoor/Outdoor Kennel

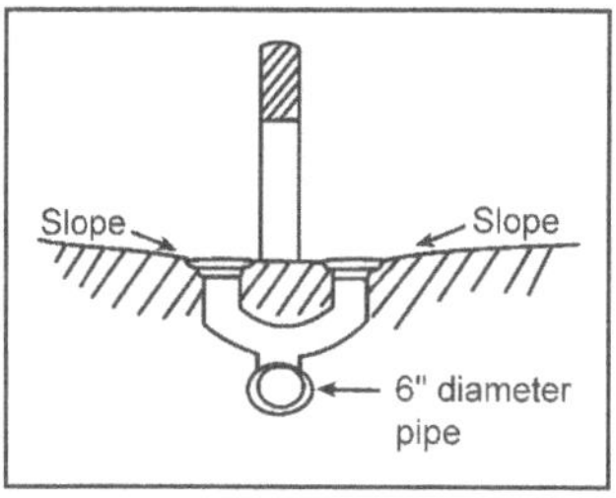

Figure B-9. Example of a Combination Indoor/Outdoor Kennel Floor Drain

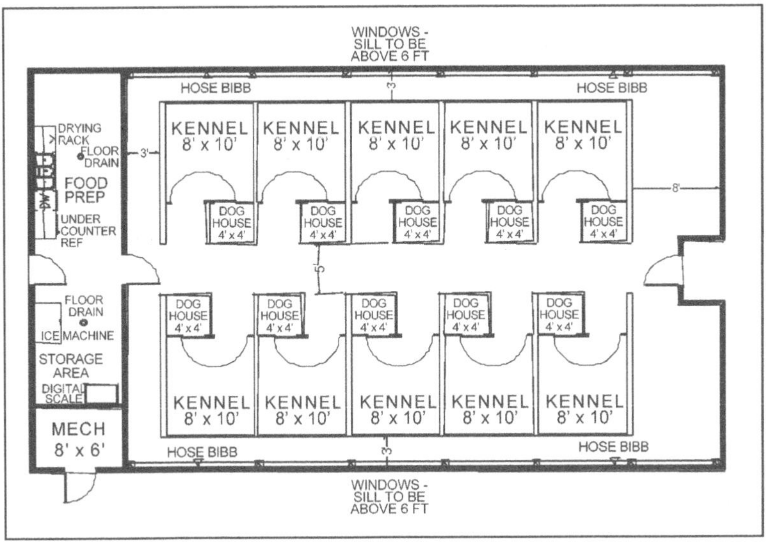

Figure B-10. Indoor or Outdoor Kennel

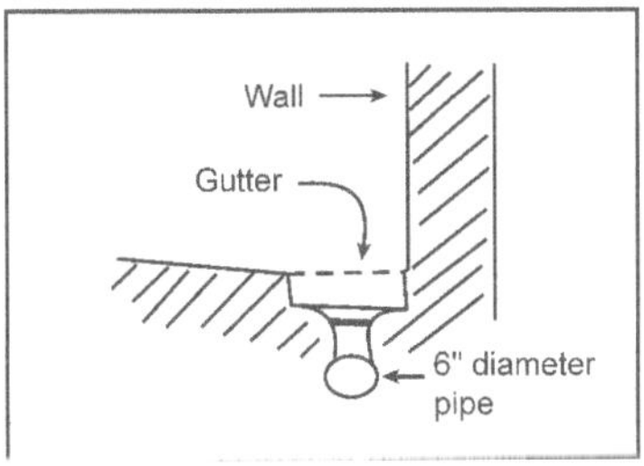

Figure B-11. Example of an Indoor or Outdoor Kennel Floor Drain

B-47. Flooring should be concrete and slope toward floor drains and/or gutters to provide rapid runoff of water and drying. Since runs are subject to frequent wash downs to clean and prevent disease, a smooth epoxy-type finish is needed. Temperature-controlled floors should be emplaced for locations that have inclement weather.

B-48. All materials in areas accessible to MWDs should be resistant to damage by scratching, biting, and chewing. Materials should be durable and easy to sanitize. Avoid using material such as angle iron that rusts over a period of time and generates sharp edges that pose a danger to MWDs.

B-49. MWD houses are 4 feet by 4 feet by 4 feet sealed, concrete boxes with a metal top. Tops should hinge toward the end wall in order to be raised to clean the house or treat an MWD. Once lifted, a holding device should be available to keep the top open. Wall opening dimensions are shown in *Figure B-12*, page B-14. The house should be a minimum of 8 inches and a maximum of 12 inches above the run floor. The house floor should have positive drainage toward the wall opening.

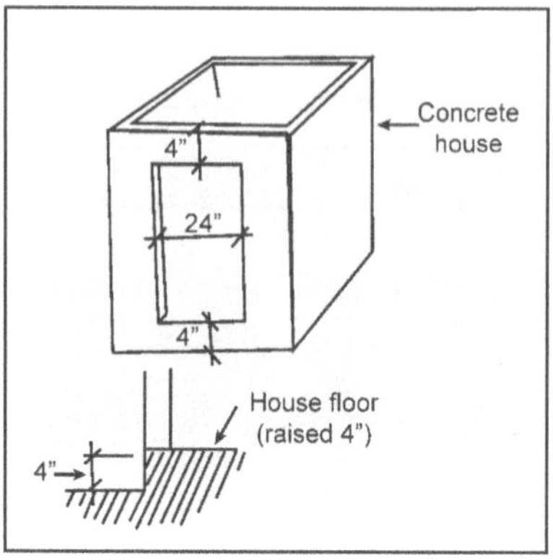

Figure B-12. MWD House

Storage and Mechanical Area

B-50. Additional storage for the kennel is provided. A commercial quality ice machine is required to aid in cooling down MWDs. Include a floor drain for the area. If a veterinary treatment room is not included, a digital scale should be located in this area. Design mechanical systems to optimize energy use and meet the design criteria for the area where the kennel is located. Include the hot water heater in the mechanical room and with a capacity to meet the demands of the kennel. Allow for the location of electrical panels in this area as well.

Food Preparation Room (112 Square Feet)

B-51. The food preparation area is where feedings are prepared and feeding pans/bowls are stored and cleaned. Food preparation for MWDs requires a large countertop with a deep, three-compartment stainless steel sink. Upper wall cabinets are required for general storage. An under the counter refrigerator, a dishwasher, and space to store food for daily use are necessary for operation. A 2-foot space at the end of the counter should accommodate a storing and drying rack for pans and bowls. Lighting and receptacles and a floor drain should be installed per the code. A telephone, bulletin board, feeding chart, and large garbage can should also be provided. Optional equipment includes a garbage disposal unit and a stackable washer and dryer.

Appendix C
Inspection Memorandum

This appendix provides a sample memorandum (see *Table C-1*, page C-2) for documenting the results of inspections. Commanders and program managers will find this useful in preparing for inspections. KMs, plans NCOs, and handlers will find this list helpful in conducting daily duties and adhering to requirements for inspections. Maintain completed memorandum records on file to assist in future inspections.

Table C-1. Sample Inspection Memorandum

DEPARTMENT OF THE ARMY
UNIT
ADDRESS OF UNIT

OFFICE SYMBOL 15 May 2004

MEMORANDUM FOR RECORD

SUBJECT: MACOM Technical Assistance Visit or Provost Marshal Inspection

1. SECTION I. GENERAL INFORMATION.

UNIT: 529th Military Police Company LOCATION: Kaiserslautern
DATE: 3-14-05 TIME: 0900

INSPECTOR (S): LTC Jeffery S. Captain ESCORT: SFC James Majors

TYPE OF VISIT: __X__ ANNOUNCED______UNANNOUNCED

TYPE	AUTHORIZED	ON HAND
Patrol		
Patrol Explosive	6	4
Patrol Narcotic	3	2
Explosive Detector	0	0
Narcotic Detector	0	0

2. SECTION II. PERSONNEL.	**YES**	**N/A**	**NO**
a) Are all authorized MWDs on hand (unit TDA/MTOE document)?			
b) Are all MWD handlers certified by the DOD Dog Center?			
c) Are ASI trained handlers being used as specified in AR 614-200?			
d) Is the assignment policy of one dog and one handler being followed?			
e) Are all MWDs assigned to a handler?			
f) Has an excess of dogs been reported to the MACOM?			
g) Does the KM meet the minimum criteria to be appointed as a KM?			
h) Are handlers being assigned duties that take away from their handling responsibilities (excluding PT, staff duty, field duty, and unit training)?			

3. RESPONSIBILITIES OF THE PROVOST MARSHAL AND COMMANDER.

	YES	N/A	NO
a) Is there an updated SOP signed by the PM explaining policies, procedures, and responsibilities for the direction, management, and control of the kennels?			
b) Has the PM initiated an MWD program based on the evaluation of the installation support, customs, HRP support, and combat support missions?			

Table C-1. Sample Inspection Memorandum (Continued)

OFFICE SYMBOL SUBJECT: MACOM Technical Assistance Visit or Provost Marshal Inspection		YES	N/A	NO
c)	Has the PM issued written guidelines governing the release of an MWD to apprehend a suspect?	____	____	____
d)	Are there specific criteria in contingency plans governing the use of dog teams during civil disturbance situations?	____	____	____
e)	Does the PM provide guidance on procedures for explosive detector dog support to civil authorities?	____	____	____
f)	Is the proficiency of each MWD team evaluated periodically by the PM?	____	____	____
g)	Has the PM and/or unit commander checked one or more of these areas monthly: kennels, kennel support building, training and exercise areas, equipment, handlers, dogs, training, utilization and team proficiency?	____	____	____
h)	Has the PM and/or unit commander inspected all areas of the MWD kennels quarterly, and is a written record on file?	____	____	____
i)	Are MWD requisitions submitted according to *AR 190-12* and *AR 700-81*?	____	____	____
j)	Are MWDs placed on the unit property book by NSN, name, and tattoo number?	____	____	____
k)	Are MWDs assigned to the KM on a *DA Form 3161* (*Request for Issue or Turn-In)* hand receipt (according to *DA Pam 710-2-1*)?	____	____	____
4.	**VCO RESPONSIBILITIES.**			
a)	Are the dogs examined by the VCO every six months (according to *AR 40-905*)?	____	____	____
b)	Does the VCO inspect the kennel facility and area for proper sanitary standards on a quarterly basis?	____	____	____
c)	Is the issued dog food and diet checked periodically?	____	____	____
d)	Is there a program in effect for insect and rodent inspection?	____	____	____
e)	Does the VCO concur with the kennel standards/facilities?	____	____	
	If not what course of action has been taken? ______________________________			
f)	Does the VCO provide regular training to handlers on MWD health care, feeding, and first aid?	____	____	____
g)	Does the VCO provide a list to the KM concerning the deployable status of all MWD?	____	____	____
h)	Does the VCO issue health certificates for dogs pending shipment?	____	____	____

Table C-1. Sample Inspection Memorandum (Continued)

OFFICE SYMBOL SUBJECT: MACOM Technical Assistance Visit or Provost Marshal Inspection	YES	N/A	NO
5. WARNING SIGNS.			
a) Are there warning signs with the following words posted at each instillation entrance, "**CAUTION—THIS AREA PATROLLED BY MILITARY WORKING DOGS**?"	___	___	___
b) Are kennels located away from disturbing influences and adequately posted with "**DANGER—OFF LIMITS—MILITARY DOG AREA**?"	___	___	___
c) Are the shipping crates stenciled with "**CAUTION—MILITARY WORKING DOG—DO NOT TAMPER WITH ANIMAL**?"	___	___	___
d) Are shipping crates stenciled with "**LIVE ANIMAL**?"	___	___	___
e) Are removable or permanent markings placed on the sides and rear of vehicles with the following phrase "**CAUTION—MILITARY WORKING DOGS**?"	___	___	___
6. KENNEL AREA.			
a) Has a risk analysis been conducted on the kennels?	___	___	___
b) Is the kennel manned both day and night?	___	___	___
c) Does the kennel have an IDS/CCTV system if not manned both day and night?	___	___	___
d) Are approved locks and padlocks used to secure government equipment?	___	___	___
e) Have primary and alternate key custodians been appointed, in writing, by the PM or commander?	___	___	___
f) Are personnel authorized to issue and receive keys listed on an access roster?	___	___	___
g) Are all keys controlled and accounted for on a *DA Form 5513-R* (*Key Control Register and Inventory)?*	___	___	___
h) Is the *DA Form 5513-R* secured in an approved, lockable container?	___	___	___
i) Is the key depository box permanently affixed to a wall?	___	___	___
j) Are keys inventoried semiannually, and is the last report on file?	___	___	___
k) Are all keys inscribed with a serial number?	___	___	___
l) Are kennels used to house or care for other animals?	___	___	___

Table C-1. Sample Inspection Memorandum (Continued)

OFFICE SYMBOL SUBJECT: MACOM Technical Assistance Visit or Provost Marshal Inspection	YES	N/A	NO
m) Are adequate training areas adjacent to the kennel areas?	____	____	____
1) Is the training area fence 8 feet high?	____	____	____
2) Does the overhanging fence face inward?	____	____	____
3) Is the fence staked to the ground?	____	____	____
n) Does the kennel area have an adequate water supply?	____	____	____
o) Are the kennels of a standard design?	____	____	____
7. EQUIPMENT.			
a) Is all authorized equipment on hand?	____	____	____
b) Are Air Force shipping crates returned within 10 days after receiving a new or replacement MWD?	____	____	____
c) Is there one shipping crate available for each dog authorized?	____	____	____
d) Is leather equipment properly maintained and cleaned?	____	____	____
e) Is all metal equipment free of rust?	____	____	____
f) Is the web equipment cleaned with proper cleaning material to prevent deterioration?	____	____	____
g) Are there adequate medical supplies on hand for first aid treatment and training?	____	____	____
8. USE OF MWDs.			
a) Are the dog teams being employed a minimum of 24 hours per week?	____	____	____
b) Are PDs employed in all areas of the installation during both day and night?	____	____	____
9. NARCOTICS.			
a) Is the MWD section registered with the Drug Enforcement Administration?	____	____	____
b) Is a *DEA 223 (Controlled Substance Registration Certificate)* on file and current?	____	____	____
c) Have all personnel been designated, in writing, by the PM to handle training aids?	____	____	____
d) Has primary and alternate narcotic training aid custodians been appointed, in writing, by the PM?	____	____	____

Table C-1. Sample Inspection Memorandum (Continued)

OFFICE SYMBOL SUBJECT: MACOM Technical Assistance Visit or Provost Marshal Inspection		YES	N/A	NO
e)	Have primary and alternate narcotic custodians been cleared by a favorable Crime Records Center name check?			
f)	Is the Crime Records Center name check maintained on file?			
g)	Have all personnel authorized to handle or use training aids been briefed at least annually?			
h)	Is a written record of the briefing being maintained on file?			
i)	Are the monthly/quarterly inventories being conducted by a disinterested E7 or above?			
j)	Are the monthly/quarterly inventories being maintained with the controlled substances?			
k)	Are the monthly/quarterly inventories being maintained for at least two years?			
l)	Are joint inventories being conducted on the change of primary or alternate custodians and is the inventory maintained on file?			
m)	Is a daily issue/return green logbook on hand?			
n)	Is the property book properly maintained to account for all authorized explosive training aids?			
o)	Is an *SF 702 (Security Container Check Sheet)* posted on the safe?			
p)	Are *SFs 702* maintained for 30 days?			
q)	Is a controlled-substance accountability folder on hand?			
r)	Is the controlled-substance accountability folder properly maintained?			
10.	**EXPLOSIVES.**			
a)	Have primary and alternate explosive custodians been appointed, in writing, by the PM?			
b)	Have primary and alternate custodians been cleared by a favorable Crime Records Center name check?			
c)	Are records of the Crime Records Center checks on file?			
d)	Are all personnel authorized to receive and handle explosive training aids designated, in writing, by PM?			
e)	Is annual explosive safety training being conducted by EOD personnel or other qualified personnel?			

Table C-1. Sample Inspection Memorandum (Continued)

OFFICE SYMBOL SUBJECT: MACOM Technical Assistance Visit or Provost Marshal Inspection		YES	N/A	NO
f)	Is a written record of the safety training being maintained on file for a minimum of two years?	____	____	____
g)	Is there an SOP that outlines safety procedures in training areas, transportation of explosives, severe weather conditions, and accidents?	____	____	____
h)	Are the monthly/quarterly inventories being conducted by a disinterested E7 or above?	____	____	____
i)	Are the monthly/quarterly inventories maintained on file for at least two years?	____	____	____
j)	Are all required explosive training aids available to ensure that proficiency standards are met?	____	____	____
k)	Are chlorates being replaced quarterly?	____	____	____
l)	Is each type of explosive being stored in a separate vapor proof container?	____	____	____
m)	Is each vapor proof container properly marked?	____	____	____
n)	Has the vehicle that is used to transport explosive training aids been approved by safety personnel?	____	____	____
o)	Does the installation have an explosive detector training plan that has been approved by the installation commander which lists buildings and facilities authorized for training the PEDD teams?	____	____	____
p)	Is a daily issue/return green logbook on hand?	____	____	____
q)	Is the logbook properly maintained to account for all authorized explosive training aids?	____	____	____
r)	Is a *DA Form 7281-R (Command Oriented Arms, Ammunition, & Explosives Security Screening and Evaluation Record)* on hand for each handler that is authorized to handle or receive explosive training aids?	____	____	____
11.	**TRAINING.**			
a)	Does the KM (or handlers) give training to military police on MWD team functions and operational considerations?	____	____	____
b)	Are the MWDs receiving a minimum of 4 hours of patrol and 4 hours of narcotic/explosives proficiency training each week?	____	____	____
c)	Have all certified narcotic/explosive detector teams received quarterly validation testing by the KM?	____	____	____
d)	Is the minimum proficiency standard for assigned patrol/narcotic detector dogs 90 percent and patrol/explosive detector dogs 95 percent?	____	____	____
e)	Was the dog decertified when his proficiency standard was less than 90 percent and less than 95 percent for more than three consecutive months?	____	____	____

Table C-1. Sample Inspection Memorandum (Continued)

OFFICE SYMBOL

SUBJECT: MACOM Technical Assistance Visit or Provost Marshal Inspection	**YES**	**N/A**	**NO**
12. RECORDS.			
a) Is the handler's information being recorded on the back of *DD Form 1834*?	____	____	____
b) Have all *DA Forms 2807-R* been completed properly?	____	____	____
c) Have all deficiencies for the MWDs been annotated and explained on DA Forms 2807-R?	____	____	____
d) Is a *DA 3992-R* being maintained to record the MWD's reliability?	____	____	____
e) Does each MWD team have a probable cause folder?	____	____	____

13. THE RESULTS OF THIS INSPECTION ARE—

____ EXCELLENT

____ SATISFACTORY

____ UNSATISFACTORY

14. REMARKS. All areas noted with a deficiency will be annotated in the remarks section. A copy of this technical assistance visit will be provided to OPMG, the MACOM program manager, and the KM. All records of technical assistance visits will be maintained on file at the installation for a minimum of two years after the inspection date with the corrective action memorandum.

15. POINT OF CONTACT. The point of contact for this action is the undersigned at COMM: 555-555-1511 or e-mail address.

(Original Signed)
JEFFREY S. CAPTAIN
LTC, USA
MACOM PM

Appendix D
Training Records

This appendix provides samples and details of how to properly use MWD training records. Records must be maintained for the life of the MWD. At least one year of historical documents must be maintained on file at each kennel for every MWD. Records can be discarded two years after the death, adoption, or euthanasia of the MWD. Nearly all MWDs have a dual purpose (PD and PEDD or PNDD); therefore, every MWD should have a PD and a PEDD/PNDD form on file. Kennels that have internet access can obtain records via the Military Working Dog Management System (MWDMS).

DEPARTMENT OF THE ARMY *FORM 2807-R*

D-1. *DA Form 2807-R (Military Working Dog Training and Utilization Record)* (*Figure D-1,* page D-3) is a standard form used by all MWD personnel. It is used to record the training and utilization of PDs. Use the following instructions when preparing a *DA Form 2807-R*:

ADMINISTRATIVE DATA

D-2. Enter the following information in the appropriate blocks:

- The date.
- The MWD's name and type, tattoo number, and age.
- The handler's name and grade.
- The organization and the location.

TRAINING

D-3. The training section of the form provides complete information on all of the training tasks that must be accomplished for a PD to maintain proficiency. It is not necessary to train on all of the controlled-aggression tasks during the same day; however, all of the controlled-aggression tasks must be trained every week.

D-4. Entries for the amount of time spent training on specific tasks during a particular day of the month are made in minutes. The monthly total for each specific task will be in hours and will be automatically entered in the far right-hand column titled "Total Hours." The handler's or trainer's evaluation of the MWD's daily rating is entered as S or U. Remarks are mandatory for any area the MWD fails to perform to standard. These remarks will be annotated on the reverse side of the form using the following format:

- Deficiency (DF). The MWD's shortcoming.
- Corrective action (CA). The handler's response to the MWD's deficiency.
- Corrective response (CR). The MWD's response to the CA.

D-5. Enter the appropriate information on—

- **Line 1.** Record the number of minutes per day that the MWD is trained on obedience on leash.
- **Line 2.** Record the number of minutes per day that the MWD is trained on obedience off leash.
- **Line 3.** Record the number of minutes per day that the MWD is trained on the obedience course.

- **Line 4.** Record the daily rating of the MWD on controlled aggression (either S or U). Write an explanation of any U rating on the reverse side of the form. In 4a to 4e, record the number of minutes per day that the MWD trained on each controlled-aggression task.
- **Line 5.** Record the number of minutes per day that the MWD trained on building searches on line 5.
- **Line 6.** Record the number of minutes per day that the MWD trained on gunfire with the handler firing the weapon. The number of rounds will be annotated on the reverse side of the form.
- **Line 7.** Record the number of minutes per day that the MWD is trained on gunfire with a third party firing a weapon. The number of rounds will be annotated on the reverse side of the form.
- **Line 8.** Record the number of minutes per day that the MWD is trained on scouting/patrolling. Record the distance (in feet) for scent, sight, and sound detection on lines 8a, b, or c.
- **Line 9.** Record the number of minutes per day that the MWD is trained on vehicle patrol.
- **Line 10.** Record the number of minutes per day that the MWD is trained on tracking.
- **Line 11.** Record the daily training rating (S or U). Explain deficiencies on the reveres side of the form. Any failure of a critical task requires a U daily training rating. See *DA Pam 190-12* for critical tasks.

UTILIZATION

D-6. The utilization section of the form provides a daily record of the time spent performing military police duties in the three general categories of combat support operations, patrol–law enforcement, and patrol–security.

D-7. The combat support operations category may be used to record time spent performing military police missions (actual or training) in support of combat units. Such operations include field training exercises, CP exercises, mobilization exercises, and other activities related to the tactical and strategic missions of the Army, provided that these missions are accomplished with a handler and an MWD together as a certified team.

D-8. The patrol–law enforcement category may be used to record time spent performing law enforcement or force protection type missions. These missions include but are not limited to—

- Random vehicle searches.
- Law enforcement patrols.
- Other duties performed while working in a law enforcement type mission.

D-9. The patrol–security category may be used to record time spent performing security type missions such as—

- Dignitary security.
- Critical asset patrolling in a peacetime environment.
- USSS, Department of State, or other explosive type searches.

D-10. The daily rating of the MWD's performance on military police duties is recorded as S or U. All U performance annotations will have an explanation for the DF and have the CA taken recorded on the reverse side of the form and continuation sheets, if necessary. The total monthly utilization hours are recorded in the last column, titled "Total Hours."

D-11. When training is conducted during the actual use and employment of an MWD, the time used for training may also be reported in the training section. This training is not an automatic entry every time the team has utilization hours. Time annotated as training during utilization hours will be actual time training on patrol or detector tasks. It is unethical to count all or even half of an MWD team's utilization hours as both utilization and training time for every shift, and unrealistic numbers will not be accepted. When time is recorded as utilization and training, a notation is made on the reverse side of the form explaining the double time entry and the type of training conducted. This will make it possible to differentiate between training time, training time while on duty, and duty (utilization) time. Although additional training during utilization hours is acceptable, it will not be accepted as the primary amount of mandatory training hours each week.

MILITARY WORKING DOG TRAINING AND UTILIZATION RECORD For use of this form, see AR 190-12; proponent agency is ODCSOPS						MONTH AND YEAR SEP 2003
NAME/TYPE OF DOG NATZ/PEDD	TATTOO/NUMBER E309	AGE 2	NAME OF HANDLER BOOKSTEIN, DAVID B.	GRADE SGT	ORGANIZATION AND LOCATION HHC 98TH ASG, BAMBERG, GE	

DAILY RATINGS: S - SATISFACTORY, U - UNSATISFACTORY (Explain deficiency and corrective action on reverse)
(Use reverse side of this form for any remarks or notes)

TRAINING	DAY OF MONTH																															TOTAL HOURS
	1	2	3	4	5	6	7	8	9	10	11	12	13	14	15	16	17	18	19	20	21	22	23	24	25	26	27	28	29	30	31	
1. ON LEASH OBEDIENCE	15	15	15	15	30		30	40	10	15			20	15		15	20	30	15		15	15	15	15					15	15		6.50
2. OFF LEASH OBEDIENCE	20	20	15	15	40		30	40	10	15			20	30		15	30	30	15		15	20	20	20					15	15		7.50
3. OBEDIENCE COURSE					20		20			15																						0.92
4. CONTROLLED AGGRESSION (S OR U)		S	S					S	S				S	S																		
a. FALSE RUN		5	5					15	5				5	5																		0.67
b. ATTACK		10	10					20	10				10	10																		1.17
c. SEARCH AND ATTACK		5	5					15	10				10	10																		0.92
d. STAND OFF		10	15					20	5				10	10																		1.17
e. ESCORT		5	5					15	5				5	5																		0.67
5. BUILDING SEARCH		15						15																20								0.83
6. GUNFIRE - HANDLER								10																								0.17
7. GUNFIRE - DECOY/AGITATOR					10														30													0.67
8. SCOUTING/PATROLLING (TIME)					15																			15					15			0.75
a. SCENT DETECTION (DISTANCE)					40																			30					30			
b. SIGHT DETECTION (DISTANCE)																																
c. SOUND DETECTION (DISTANCE)																																
9. VEHICLE PATROL																																0.00
10. TRACKING																																0.00
11. DAILY TRAINING RATING (S OR U)	S	S	S	S	S		S	S	S	S						S	S	S	S		S	S	S	S					S	S		21.92

UTILIZATION	HOURS PER DAY																															TOTAL HOURS
	1	2	3	4	5	6	7	8	9	10	11	12	13	14	15	16	17	18	19	20	21	22	23	24	25	26	27	28	29	30	31	
1. COMBAT SUPPORT OPERATIONS																																0
2. PATROL - LAW ENFORCEMENT	8	8	8	8	8			8	8	8			8	8	8	8				8	8	8	8						8	8		144
3. PATROL - SECURITY																																0
4. DAILY UTILIZATION RATING (S OR U)	S	S	S	S	S			S	S	S			S	S	S	S				S	S	S	S						S	S		144.0

DAILY FEEDING	QUANTITY OF FOOD (BY WEIGHT)																															WT OF DOG	
	1	2	3	4	5	6	7	8	9	10	11	12	13	14	15	16	17	18	19	20	21	22	23	24	25	26	27	28	29	30	31	DATE	LBS
2 CU S/D	A	A	A	A	A	A	A	A	A	A	A	A	A	A	A	A	A	A	A	A	A	A	A	A	A	A	A	A	A	A	A	1ST	75
2 CU S/D	A	A	A	A	A	A	A	A	A	A	A	A	A	A	A	A	A	A	A	A	A	A	A	A	A	A	A	A	A	A	A	15TH	73.5

DA FORM 2807-R, Oct 84 EDITION DEC 72 IS OBSOLETE, AND REPLACES DA FORMS 2810-R AND 2815-R, DEC 72

Sample

Figure D-1. Sample DA Form 2807-R

DA Form 2807-R
Continuation Sheet

DATE	REMARKS
2-Sep-03	**DF1:** MWD WAS DISTRACTED BY OTHER HANDLER; WOULD NOT FOCUS ON COMMANDS.
	CA1: CONDUCTED PROPER AVOIDANCE TRAINING TECHNIQUE.
	CR1: AFTER MWD DID NOT RESPOND TO INITIAL AVOIDANCE TRAINING TECHNIQUE, A PINCH COLLAR WAS USED, RESULTING IN REGAINED FOCUS.
	DF2: MWD BIT SLEEVE ON STAND OFF.
	CA2: CONDUCTED PROPER AVOIDANCE TRAINING TECHNIQUE.
	CR2: MWD PERFORMED STAND OFF CORRECTLY.
5-Sep-03	GUNFIRE TRAINING: 5 ROUNDS FIRED BY THIRD PARTY.
	DF1: MWD REPEATEDLY BROKE POSITION (SIT TO DOWN AND DOWN TO SIT).
	CA1: CONDUCTED PROPER AVOIDANCE TRAINING TECHNIQUE.
	CR1: MWD REMAINED IN POSITION.
	DF2: MWD WAS HESITANT TO OUT REWARD (BITE SLEEVE).
	CA2: CONDUCTED PROPER AVOIDANCE TRAINING TECHNIQUE.
	CR: MWD OUTED REWARD ON COMMAND.
7-Sep-03	MWD HAS BEEN BREAKING POSITION AFTER SHORT TIME DURING OFF/L OB WHEN HANDLER IS POSITIONED AWAY FROM HIM. TODAY MWD REMAINED IN THE SIT FOR 1 MIN 50 SEC, BROKE SIT, WAS VERBALLY CORRECTED AND REMAINED IN SIT FOR AN ADDITIONAL 25 SEC.
	DF: MWD BROKE FROM DOWN SEVERAL TIMES DURING PARADE REST.
	CA: CONDUCTED PROPER AVOIDANCE TRAINING TECHNIQUE.
8-Sep-03	GUNFIRE TRAINING: 10 ROUNDS FIRED BY HANDLER.
	CR: AFTER MWD DID NOT RESPOND TO AVOIDANCE TRAINING, ESCAPE TRAINING WAS CONDUCTED. EVENTUALLY MWD REMAINED IN DOWN. MWD IS SO EXCITABLE THAT HE HAS A TENDENCY TO CUE OFF HANDLER'S MOVEMENTS TO BREAK FROM POSITION. HANDLER IS CONCENTRATING ON TEACHING MWD NOT TO BE SO DISTRACTED BY OUTSIDE STIMULUS.
	DF1: MWD WAS GETTING DISTRACTED AND NOT HEELING IN THE PROPER POSITION.
	CA1: CONDUCTED PROPER AVOIDANCE TRAINING TECHNIQUE (HEEL-A-WAYS).
	CR1: MWD HEELS CORRECTLY.
	DF2: MWD WAS BITING SLEEVE DURING STAND OFF.
11-Sep-03	**CA2:** CONDUCTED PROPER AVOIDANCE TRAINING TECHNIQUE.
	CR2: MWD PERFORMED STAND OFF CORRECTLY.
	NO TRAINING DUE TO MWD GOING TO VET. MWD HAS BEEN DIAGNOSED WITH A POSSIBLE PULLED/STRAINED MUSCLE IN HIS RIGHT LEG, WHICH HE IS FAVORING AS HE WALKS. X-RAYS SHOW NO BREAKS OR FRACTURES. MWD CANNOT DO ANYTHING STRENUOUS FOR THE NEXT WEEK (NO RUNNING OR JUMPING UNTIL 18 SEP) INCLUDING BITE WORK OR OB COURSE. MWD CAN DO BASIC OB STARTING TUES, SEP 16.
19-Sep-03	GUNFIRE TRAINING: 15 ROUNDS FIRED BY DECOY.

Figure D-1. Sample DA Form 2807-R (Continued)

D-12. Record the appropriate utilization information on—

- **Line 1.** Record the hours per day that the MWD is used for combat support operations.
- **Line 2.** Record the hours per day that the MWD is used for patrol–law enforcement.
- **Line 3.** Record the hours per day that the MWD is used for patrol–security.
- **Line 4.** Record the daily utilization rating (S or U). Explain any U on the reverse side of the form.

TOTAL HOURS

D-13. Record the total hours of training and use in the last column on the right side of the form. The time should be recorded in hours and minutes. The MWDMS makes this entry automatically.

DAILY FEEDING

D-14. Record the quantity of food that the MWD eats every day in this section. If the MWD is fed once daily, the type and amount of food will be noted in the first block to the left. Under quantity of food, the amount eaten for each feeding will be annotated as follows:

- "A" if all food was eaten.
- "3/4" if three-quarters was eaten.
- "1/2" if half was eaten.
- "1/4" if one-quarter was eaten.
- "0" if none of the food was eaten.

D-15. If the MWD is fed twice daily, the second block is used to record the second daily feeding (which could be a different type of food). If the MWD is fed once daily, the second block may be used to record a change in the diet of the MWD, directed by the veterinarian. Block 2 may also be used to indicate the type and amount of food used daily for MWDs on a food reward system. Monitoring MWD food consumption is a critical portion of the required daily health checks.

D-16. The last column, titled "Wt of Dog," is used to record a semimonthly weight for the MWD. The weight checks occur on or around the first and the fifteenth day of each month. There is sufficient room under the headings "Date" and "Lbs" to record both the date when the MWD is weighed and the weight in pounds. Any drastic changes in food consumption or weight gain or loss are reported to the veterinarian.

D-17. Enter the following information in the appropriate blocks:

- Enter the VCO's prescribed amount of food to be given daily directly under "Daily Feeding." Both blocks are used when an MWD is fed twice daily.
- Record the amount of food eaten per day, in the appropriate day's block under "Quantity of Food (by Weight)."
- Record the date and weight of the MWD in the respective blocks in the bottom right-hand corner.

REMARKS

D-18. The reverse side of *DA Form 2807-R* (*Figure D-1*) is used to annotate advanced techniques used during training and DFs with CAs and CRs listed. This section is very important for probable cause and shows the time line of events for training. Each day is listed with the action taken on that day. If no training is conducted due to company type training, the words "soldier development" are written on the line followed by the type of training conducted. "Soldier development" is only used when MWD handlers are conducting training in which the MWD is not needed or not taken. If no training is conducted due to leave, passes, or scheduled days off, no entry is needed; *DA Form 6 (Duty Roster)* can be used to verify these days.

DEPARTMENT OF THE ARMY *FORM 3992-R* (FIRST PAGE)

D-19. *DA Form 3992-R (Narcotics or Explosives Detector Dog Training and Utilization Record)* (*Figure D-2*, page D 9) is a standard form used by all MWD personnel. It is used to record the use and training of detector dogs. Use the following instructions when preparing a *DA Form 3992-R:*

ADMINISTRATIVE DATA

D-20. Enter the following information in the appropriate blocks:

- The date.
- The MWD's name and type, tattoo number, and age.
- The handler's name and grade.
- The organization and location.

TRAINING

D-21. The training section of the form is used for recording all training of the detector dog team in the particular detection skill. The detection areas of buildings, containers, and vehicles are listed. There is one blank space for detector training in other areas. Examples of other areas may include—

- Mail facilities.
- Open or wooded areas.
- Warehouses.
- Aircraft.
- Areas considered significant or unique to the command to which the MWD team is assigned.

D-22. Although some of the examples may be included under one of the three prelisted categories, the others should be listed separately. For areas covered under the three general headings, the handler or trainer annotates the specific location under the appropriate day in the remarks (for example, Smith High School, NCO Club, or Corps Auditorium).

D-23. The top line of each training category is divided diagonally into two blocks. The top of each block is used to record the number of training aids planted for that day's detector training. The bottom of each block is used for the number of training aids found by the proper response of the detector dog. If the MWD responds on the same training aid more than once in the same training scenario, it will not be counted as a find unless the number of plants is changed as well. At no time will there be more finds annotated than there are plants. Continually allowing the MWD to return to the last aid found creates false response problems in the MWD and should not be encouraged. It is the handler's responsibility to keep the MWD from returning to the scent cone of the last found training aid, whether the MWD is being worked on or off leash. Once the MWD has entered the scent cone, he must be allowed to work to the source and final respond. At no time will an MWD be yanked off an aid once he is working the scent cone.

D-24. The bottom line of each training category is used to record the search time. Search time is the amount of time devoted to detector training that day. Detector training is the time from the moment the handler enters a problem and commands the MWD to "Seek" until the MWD has completed the search of that training problem, excluding handler or MWD breaks. Add the daily time entries in each training category, and enter the results in the last column, titled "Total Hours."

D-25. Record the following information beside the appropriate training location (building, containers, vehicles, or other):

- **Plants/Finds.** Record the number of plants in the top half of the block and the number of finds in the lower half.
- **Search Time.** Record the amount of time devoted for detector training that day.
- **Training Total Hours.** Record the sum of hours performing detection missions in the far right

UTILIZATION

D-26. All operational detection missions are recorded in the utilization section. The detection missions are listed. Detection missions that are significantly different from these are listed separately to give an accurate record of the types of missions for which the detector dog team is used. There are two blocks under each day of the month for each utilization category. The top block is for recording the number of times the MWD detected a substance. The bottom block is for recording the total search or inspection time for the detection mission. The last column is the sum of the daily times. It gives a record of the total amount of time spent performing the detection missions for that month. The number of finds for the month can also be totaled and entered in the last column, above the time entry.

D-27. Record the following information beside the appropriate location (building, containers, vehicles, or other):

- **Finds.** Record the number of times the MWD finds the substance on a detection mission.
- **Search Time.** Record the total search time.

DETECTOR DOG SEARCH DATA

D-28. All relevant information about the productivity and success of each detection mission is recorded using the "Detector Dog Search Data" section. The location may be a building number, a unit designation, map grid coordinates, and/or any other information that helps identify where the detection mission was performed.

D-29. The MPR number is the number assigned to the case by military police to account for the custody and disposition of the substances found. The substance is the identification of the found material by a common name, such as heroin, marijuana, dynamite, or detonating cord.

D-30. Quantity is the measure of the amount of the substance found. Weight, volume, overall dimensions, length, or any other appropriate measure may be used.

D-31. The "Remarks" section may be used for adding any other relevant information about the substance found. This would include a field or laboratory verification of the type of substance, an EOD evaluation of an explosive device or explosive substance as live or inert, or the presence of other hazardous material in or around the substance found (such as razor blades, trip wires, or poisons). Record any information that may be useful in preparing the team for future searches or that may be applicable or useful to other detector dog teams.

D-32. Record the following information in the appropriate boxes:

- **Time.** Record the time the detector dog performed the mission.
- **Date.** Record the date the detector dog performed the mission.
- **Location.** Record the exact location of the detector mission (such as front gate, Fort Sill, Oklahoma; National Airport, Gabon, Africa; or building 442, Fort Bragg, North Carolina).
- **Military Police Report (MPR) Number.** Record the MPR number in this column, if applicable.
- **Substance.** Record the common name of the substance found, if applicable.
- **Quantity.** Record the quantity of the substance found, if applicable.
- **Remarks.** Use this column to record any other relevant information about the substance found (such as a positive response, no response, or a lab verification).

DETECTOR DOG PROFICIENCY (SECOND PAGE)

D-33. A correct response occurs when the MWD detects the substance and responds with the proper final response, and the training aid or substance is found in the general location where the MWD has responded. A not-at-source response is a training tool used to help an MWD team move closer to the source location of a training aid plant. The distance away from the source in order for a not-at-source response to be given is six to ten feet.

D-34. A missed response is when the MWD fails to detect and respond to the presence of a training aid with a final response. A handler miss or error is when—

- The handler fails to search or clear an area with the MWD and a training aid plant is missed.
- The handler fails to read an MWD's change of behavior and pulls the MWD away from the area where a training aid is planted and a training aid plant is missed.
- The handler presents an area but fails to ensure that the MWD is sniffing and a training aid plant is missed.

Handler misses or errors are considered missed explosives or narcotics and are counted against the find rate percentages of the MWD team.

D-35. False responses are noted in the second to last section of *DA Form 3992-R*. A false response is when the MWD responds as if he has detected a substance and the handler accepts that response, but no training aid or substance can be found within a reasonable distance of the MWD's final response.

D-36. The detector dog team's proficiency is computed monthly by adding the total number of correct responses ("a" for the month) and adding the total number of false/missed (MWD or handler) responses ("b" for the month). These two numbers are used in the following formula to obtain the detector dog team's proficiency rating:

$$\left(\frac{a}{a+b}\right)100 = proficiency$$

D-37. For example, during the month the detector dog team made 93 correct responses on training aids, the MWD team also had 4 false responses, 2 MWD and 1 handler missed responses for a total of 7 false/missed responses. Applying these two numbers (a=93 and b=7) to the formula, the following is obtained:

$$\left(\frac{93}{93+7}\right)100 = 93\%$$

D-38. The computation above shows that the MWD team is working at a 93 percent proficiency rate. This rate is above the minimum for a PNDD team (90 percent), but it is below the minimum standard for a PEDD team (95 percent). A PNDD handler should continue training to maintain and possibly increase the team's proficiency. A PEDD handler needs to identify the causes of the team's substandard performance and immediately begin corrective training to bring the team up to and over the minimum 95 percent detection proficiency standard.

> ***Note:*** Responses on actual substances during actual searches are not included in this computation. Search data cannot be included because it is impossible to determine the number of false or missed responses during an actual search.

D-39. Record the following information on—

- **Line a.** Record the total correct alerts. The MWDMS makes these entries automatically.
- **Line b.** Record the total false or missed alerts. The MWDMS makes these entries automatically.

Narcotics or Explosive Detector Dog Training and Utilization Record (Third and Fourth Pages)

D-40. The first part of section of the Narcotics or Explosive Detector Dog Training and Utilization Record section is used to record the number of each training aid used in training on each day of the month. The entry for each training aid type is divided into three rows to record the day of the month, correct alerts, and missed alerts, respectively. Record the appropriate information in the following sections.

- **Administrative Data.** Record the MWD's name and tattoo number, the month, and the year in the appropriate blocks.
- **Aid Type.** Enter the specific aid type in the upper left-hand corner of each "Detector Dog Proficiency" section.
- **Total Correct Alerts.** Record the number of correct alerts the MWD made during training.
- **Missed Alerts.** Record the number of missed and false alerts the MWD made during training.
- **Proficiency.** Record the actual proficiency in this block. The MWDMS makes this entry automatically.

Remarks

D-41. Remarks for *DA Form 3992-R* are recorded on a continuation sheet. Each DF (for example, a missed aid or false sit) is listed by the date it occurred. The CA and CR are also listed. If more than one DF is noted per day, number the DF, CA, and CR series to avoid confusion.

NARCOTICS OR EXPLOSIVES DETECTOR DOG TRAINING AND UTILIZATION RECORD

For use of this form see AR 190-12. The proponent is DCSPER.

MONTH AND YEAR	NAME/TYPE OF DOG	TATTOO NUMBER	AGE	NAME OF HANDLER	GRADE	ORGANIZATION AND LOCATION
SEP 2003	NATZ/PEDD	E309	2	BOOKSTEIN, DAVID B.	SGT	HHC 98TH ASG, BAMBERG, GE

TRAINING		1	2	3	4	5	6	7	8	9	10	11	12	13	14	15	16	17	18	19	20	21	22	23	24	25	26	27	28	29	30	31	TRAINING TOTAL HOURS 24.25
BUILDINGS	PLANTS /			7	8	4				4	6						11		5						4	8	6				7		70
	FINDS			7	5	4				4	6						10		4						4	8	6				7		65
	SEARCH TIME			120	80	60				60	60						130		70						60	120	120				120		16.67 HRS
CONTAINERS	PLANTS /									19	57						19		19														114
	FINDS									19	57						19		19														114
	SEARCH TIME									45	120						45		45														4.25 HRS
VEHICLES	PLANTS /					4													5														9
	FINDS					4													5														9
	SEARCH TIME					60													80														2.33 HRS
OTHER	PLANTS /									4																							4
	FINDS									4																							4
	SEARCH TIME									60																							1.00 HRS

UTILIZATION		-	2	3	4	5	6	7	8	9	10	11	12	13	14	15	16	17	18	19	20	21	22	23	24	25	26	27	28	29	30	31	UTILIZATION TOTAL HOURS 0.00
BUILDINGS	FINDS																																0
	SEARCH TIME																																0.00 HRS
CONTAINERS	FINDS																																0
	SEARCH TIME																																0.00 HRS
VEHICLES	FINDS																																0
	SEARCH TIME																																0.00 HRS
OTHER	FINDS																																0
	SEARCH TIME																																0.00 HRS

DA FORM 3992-R, Oct 84 EDITION OF 1 DEC 72 IS OBSOLETE

Figure D-2. Sample DA Form 3992-R

DETECTOR DOG SEARCH DATA

NO.	TIME	DATE	LOCATION	MPR NUMBER	SUBSTANCE	QUANTITY	REMARKS
1							
2							
3							
4							
5							
6							
7							
8							
9							
10							
11							
12							
13							
14							
15							
16							
17							
18							
19							
20							
21							
22							
23							
24							
25							
26							
27							
28							
29							
30							
31							
32							
33							

DETECTOR DOG PROFICIENCY
(Alerts on training aids during training and actual searches)

DAY OF MONTH	1	2	3	4	5	6	7	8	9	10	11	12	13	14	15	16	17	18	19	20	21	22	23	24	25	26	27	28	29	30	31	PROFICIENCY = a / (a + b) X 100
a. TOTAL CORRECT ALERTS	0	0	7	5	8	0	0	0	26	63	0	0	0	0	0	29	0	28	0	0	0	0	0	4	8	6	0	0	0	7	0	191
b. TOTAL FALSE/MISSED ALERTS	0	0	0	3	0	0	0	0	1	0	0	0	0	0	0	3	0	1	0	0	0	0	0	0	0	0	0	0	0	0	0	199 96%

DA Form 3992-R, Oct 84

Sample

Figure D-2. Sample DA Form 3992-R (Continued)

NARCOTICS OR EXPLOSIVES DETECTOR DOG TRAINING AND UTILIZATION RECORD

NATZ/PEDD TAT¯OO: E309 MONTH: SEP YEAR: 2003

DETONATING CORD — DETECTOR DOG PROFICIENCY (ALERTS ON TRAINING AIDS DURING TRAINING AND ACTUAL SEARCHES)

DAY OF MONTH	1	2	3	4	5	6	7	8	9	10	11	12	13	14	15	16	17	18	19	20	21	22	23	24	25	26	27	28	29	30	31	PROFICIENCY A / A + B X 100
a. TOTAL CORRECT ALERTS			2	1						20						1									3							27
b. TOTAL MISSED ALERTS				1																												28 96%

C-4 PLASTIC EXPLOSIVE — DETECTOR DOG PROFICIENCY (ALERTS ON TRAINING AIDS DURING TRAINING AND ACTUAL SEARCHES)

DAY OF MONTH	1	2	3	4	5	6	7	8	9	10	11	12	13	14	15	16	17	18	19	20	21	22	23	24	25	26	27	28	29	30	31	PROFICIENCY A / A + B X 100
a. TOTAL CORRECT ALERTS				1					22	1						3		2								1						30
b. TOTAL MISSED ALERTS				1					1																							32 94%

POTASSIUM CHLORATE — DETECTOR DOG PROFICIENCY (ALERTS ON TRAINING AIDS DURING TRAINING AND ACTUAL SEARCHES)

DAY OF MONTH	1	2	3	4	5	6	7	8	9	10	11	12	13	14	15	16	17	18	19	20	21	22	23	24	25	26	27	28	29	30	31	PROFICIENCY A / A + B X 100
a. TOTAL CORRECT ALERTS									1							20		2						2	1							26
b. TOTAL MISSED ALERTS																1																27 96%

SODIUM CHLORATE — DETECTOR DOG PROFICIENCY (ALERTS ON TRAINING AIDS DURING TRAINING AND ACTUAL SEARCHES)

DAY OF MONTH	1	2	3	4	5	6	7	8	9	10	11	12	13	14	15	16	17	18	19	20	21	22	23	24	25	26	27	28	29	30	31	PROFICIENCY A / A + B X 100
a. TOTAL CORRECT ALERTS					1				2									21						1	1							26
b. TOTAL MISSED ALERTS																		1														27 96%

WATER GEL (TOVEX) — DETECTOR DOG PROFICIENCY (ALERTS ON TRAINING AIDS DURING TRAINING AND ACTUAL SEARCHES)

DAY OF MONTH	1	2	3	4	5	6	7	8	9	10	11	12	13	14	15	16	17	18	19	20	21	22	23	24	25	26	27	28	29	30	31	PROFICIENCY A / A + B X 100
a. TOTAL CORRECT ALERTS				1						21						1		1								2				2		28
b. TOTAL MISSED ALERTS				1																												29 97%

DA FORM 3992-R, OCT 84 (CONTINUATION SHEET)

Figure D-2. Sample DA Form 3992-R (Continued)

NARCOTICS OR EXPLOSIVES DETECTOR DOG TRAINING AND UTILIZATION RECORD

NATZ/PEDD TATTOO: E309 MONTH: SEP YEAR: 2003

SEMTEX DETECTOR DOG PROFICIENCY (ALERTS ON TRAINING AIDS DURING TRAINING AND ACTUAL SEARCHES)

DAY OF MONTH	1	2	3	4	5	6	7	8	9	10	11	12	13	14	15	16	17	18	19	20	21	22	23	24	25	26	27	28	29	30	31	PROFICIENCY A / (A + B) X 100
a. TOTAL CORRECT ALERTS			1		2				1							2		2							1					2		11
b. TOTAL MISSED ALERTS																																11 100%

AMMONIUM NITRATE DETECTOR DOG PROFICIENCY (ALERTS ON TRAINING AIDS DURING TRAINING AND ACTUAL SEARCHES)

DAY OF MONTH	1	2	3	4	5	6	7	8	9	10	11	12	13	14	15	16	17	18	19	20	21	22	23	24	25	26	27	28	29	30	31	PROFICIENCY A / (A + B) X 100
a. TOTAL CORRECT ALERTS																																0
b. TOTAL MISSED ALERTS																																0 #DIV/0!

NITROGLYCERIN DYNAMITE DETECTOR DOG PROFICIENCY (ALERTS ON TRAINING AIDS DURING TRAINING AND ACTUAL SEARCHES)

DAY OF MONTH	1	2	3	4	5	6	7	8	9	10	11	12	13	14	15	16	17	18	19	20	21	22	23	24	25	26	27	28	29	30	31	PROFICIENCY A / (A + B) X 100
a. TOTAL CORRECT ALERTS				2						21															1							24
b. TOTAL MISSED ALERTS																																24 100%

TNT (Trinitrotoluene) DETECTOR DOG PROFICIENCY (ALERTS ON TRAINING AIDS DURING TRAINING AND ACTUAL SEARCHES)

DAY OF MONTH	1	2	3	4	5	6	7	8	9	10	11	12	13	14	15	16	17	18	19	20	21	22	23	24	25	26	27	28	29	30	31	PROFICIENCY A / (A + B) X 100
a. TOTAL CORRECT ALERTS			3		3											2										1				2		11
b. TOTAL MISSED ALERTS																																11 100%

SMOKELESS POWDER DETECTOR DOG PROFICIENCY (ALERTS ON TRAINING AIDS DURING TRAINING AND ACTUAL SEARCHES)

DAY OF MONTH	1	2	3	4	5	6	7	8	9	10	11	12	13	14	15	16	17	18	19	20	21	22	23	24	25	26	27	28	29	30	31	PROFICIENCY A / (A + B) X 100
a. TOTAL CORRECT ALERTS			1		2																			1	1	2				1		8
b. TOTAL MISSED ALERTS																																8 100%

	1	2	3	4	5	6	7	8	9	10	11	12	13	14	15	16	17	18	19	20	21	22	23	24	25	26	27	28	29	30	31	
Totals from TNG Aids (Cont. Sheet)																																
TOTAL PLANTS	0	0	7	8	8	0	0	0	27	63	0	0	0	0	0	30	0	29	0	0	0	0	0	4	8	6	0	0	0	7	0	197
TOTAL FINDS	0	0	7	5	8	0	0	0	26	63	0	0	0	0	0	29	0	28	0	0	0	0	0	4	8	6	0	0	0	7	0	191 97%
a. TOTAL FALSE ALERTS																2																2 1%
Totals from TNG locations (3992 Front)																																
TOTAL PLANTS	0	0	7	8	8	0	0	0	27	63	0	0	0	0	0	30	0	29	0	0	0	0	0	4	8	6	0	0	0	7	0	197
TOTAL FINDS	0	0	7	5	8	0	0	0	27	63	0	0	0	0	0	29	0	28	0	0	0	0	0	4	8	6	0	0	0	7	0	192 97%

DA FORM 3992-R, OCT 84 (CONTINUATION SHEET)

Figure D-2. Sample DA Form 3992-R (Continued)

DA Form 3992-R
Continuation Sheet

DATE	REMARKS
3-Sep-03	TRAINING AT SCHWEINFURT BOWLING ALLEY. (TRAINING AIDS PLANTED FROM 6 FT H x 5 FT D)
4-Sep-03	TRAINING AT SCHWEINFURT FMO. (TRAINING AIDS PLANTED FROM 8 FT H x 3 FT D)
	DF1: MWD MISSED AID (1 STRAND DET CORD - 1 FT H x 1 FT D) DUE TO HANDLER ERROR.
	CA1: HANDLER TOLD TO WORK A REVERSE PATTERN UNTIL THE TEAM REACHED THE AID LOCATION.
	CR1: MWD FOUND AID AFTER HANDLER MADE PROPER PRESENTATION.
	DF2: MWD MISSED AID. (1/4 BLOCK C4 2 FT H x 1 FT D)
	CA2: MWD BROUGHT BACK FOR RE-PRESENTATION.
	CR2: MWD FOUND AID AFTER HANDLER MADE SECOND PRESENTATION.
	DF3: MWD MISSED AID. (1 STICK NITRO DYNAMITE- 7 FT H x 1 FT D)
	CA3: MWD BROUGHT BACK FOR RE-PRESENTATION.
	CR3: MWD FOUND AID AFTER HANDLER MADE HIGH PRESENTATION AND DRAGGED HAND UPWARD.
	DF4: MWD MISSED AID. (1 STICK WATER GEL - 5 FT H x 1 FT D)
	CA4: MWD BROUGHT BACK FOR RE-PRESENTATION.
	CR4: MWD FOUND AID AFTER HANDLER MADE SECOND PRESENTATION.
	MWD WILL BE PROTOCOLED ON ALL MISSED AIDS AT A LATER DATE DUE TO TIME CONSTRAINTS.
5-Sep-03	TRAINING AT 630TH MILITARY POLICE COMPANY MOTOR POOL. (TRAINING AIDS PLANTED FROM 3 FT H x 7 FT D)
9-Sep-03	TRAINING AT SCHWEINFURT OPEN AREA BY CONN CLUB & OUTDOOR REC. (TRAINING AIDS PLANTED FROM 7 FT H x 4 FT D)
	CA: (REFERENCE DF2, 4 SEP) MWD WAS PROTOCOLED ON 1/2 BLOCK C4.
	CR: COMPLETED 4 HV ASSISTED FOR 4 TRIALS AND 15 CONSECUTIVE UNASSISTED TRIALS.
10-Sep-03	TRAINING AT ANSBACH CMFO WAREHOUSES (TRAINING AIDS PLANTED FROM 4 FT H x 1 FT D).
	CA: (REFERENCE DF2-4, 4 SEP) MWD WAS PROTOCOLED ON 1 STRAND DC, 1 STICK ND, AND 1 STICK WG.
	CR: COMPLETED 4 HV ASSISTED FOR 4 TRIALS AND 15 CONSECUTIVE UNASSISTED TRIALS FOR EACH AID.
11-Sep-03	NO TRAINING DUE TO MWD GOING TO VET. MWD HAS BEEN DIAGNOSED WITH A POSSIBLE PULLED/STRAINED MUSCLE IN HIS RIGHT LEG WHICH HE IS FAVORING AS HE WALKS. X-RAYS SHOW NO BREAKS OR FRACTURES. MWD CANNOT DO ANYTHING STRENUOUS FOR THE NEXT WEEK, BUT IS CLEARED TO WORK DETECTION NEXT TUES (16 SEP).
16-Sep-03	TRAINING AT KITZIGEN TISA WAREHOUSE & LEIGHTON THEATER. (TRAINING AIDS PLANTED FROM 9 FT H x 6 FT D.)
	DF1: MWD MISSED AID. (PC - 1 FT H x 6 FT D)
	CA1: MWD BROUGHT BACK FOR REPRESENTATION. WILL CONDUCT PROTOCOL.
	CR1: MWD RESPONDED ON AID AFTER HANDLER MADE SECOND PRESENTATION. MWD WAS PROTOCOLED ON PC COMPLETING 4 HV ASSISTED FOR 4 TRIALS AND 15 CONSECUTIVE UNASSISTED TRIALS.
	DF2: MWD FALSE SAT IN THEATER LOBBY.
	CA2: MWD WAS ESCAPED FROM AREA AND CONTINUED TO WORK.
	CR2: MWD CONTINUED PATTERN WITHOUT FALSE SITTING.
	DF3: MWD FALSE SAT IN THEATER BACK WALL.
	CA3: MWD WAS ESCAPED FROM AREA AND CONTINUED TO WORK. MWD WILL BE WORKED ON BLANK PROBLEM.
	CR3: MWD CONTINUED PATTERN WITHOUT FALSE SITTING.
18-Sep-03	TRAINING AT SCHWEINFURT STRIP LOT & OUTDOOR REC. (TRAINING AIDS PLANTED FROM 8 FT H x 6 FT D)
	CA1: (REFERENCE DF2, 16 SEP) CONDUCTED BLANK PROBLEM.
	CR1: MWD DID NOT ATTEMPT TO FALSE SIT.
	DF2: MWD MISSED AID. (SC - 7 FT H x 3 FT D)
	CA2: MWD BROUGHT BACK FOR RE-PRESENTATION. MWD WAS PROTOCOLED ON SC.
	CR2: MWD FOUND AID AFTER HANDLER MADE SECOND PRESENTATION. COMPLETED 4 HV ASSISTED FOR 4 TRIALS AND 15 CONSECUTIVE UNASSISTED TRIALS.
24-Sep-03	TRAINING AT SCHWEINFURT FMO. (TRAINING AIDS PLANTED FROM 5 FT H x 5 FT D)
25-Sep-03	TRAINING AT THEATER. (TRAINING AIDS PLANTED FROM 2 FT H x 2 FT D)
26-Sep-03	TRAINING AT COMMISSARY. (TRAINING AIDS PLANTED FROM 9 FT H x 6 FT D)
30-Sep-03	TRAINING AT FURNITURE WAREHOUSE. (TRAINING AIDS PLANTED FROM 6 FT H x 7 FT D)

Figure D-2. Sample DA Form 3992-R (Continued)

Appendix E

Deployment Equipment List

Table E-1 is a list to be used as a general guide to aid leaders in the planning process prior to MWD deployments. Mission, enemy, terrain, troops, time available, and civilian considerations (METT-TC) should be considered when determining what equipment is necessary for mission success. Required and authorized equipment is determined by tables of organization and equipment (TOEs), modification tables of organization and equipment (MTOEs), and tables of distribution and allowances (TDAs).

Table E-1. Deployment Equipment List

	Worn or Carried	
M9 with LCE (LBV)	Identification card	Body armor
Canteen cup	Identification tag	Black gloves with liners
First aid pouch with dressing	Protective mask with a carrier	Military police brassard (subdued)
Kevlar® helmet		
	MWD Equipment	
Choke chain	Grooming brush/comb	Feed pan
Leather or nylon collar	Reward ball/Kong	Kennel crate
Stake-out chain	360-inch nylon leash	16-foot leash
Plastic or leather muzzle	Water bucket	Harness
	Alice Pack	
BDU (2 sets)	Towels (2)	Sleeping bag with a cover
Brown T-shirt (2)	Gloves, black, cold weather	Shower shoes
Undergarments (2)	BDU Cap	PT uniform (complete set)
Green and black socks (2)	Boot polish kit	MOPP gear in an NBC bag
Poncho with a liner	Flashlight	Boots
Shaving kit	Lip balm	Personal hygiene gear
Waterproof bag	Foot powder	
	A/B Bag	
Duffel bag with a padlock	Brown T-shirt (5)	White socks (3)
Cold weather boots	BDU (2)	Green and black socks (5)
Cold weather gear	Towels (3)	Personal hygiene gear
Waterproof bag	Overboots	Undergarments (5)
Military police brassard	Laundry bag (2)	
	Additional Equipment	
30-90 day supply MWD food	50-gallon trash can (MWD food)	5-gallon water can
MWD training records	Medical supplies	Engineer tape
MWD medical records	Carpenter kit/tools	Cots
MWD health certificate	Safe for narcotics aides	Fire extinguishers

Table E-1. Deployment Equipment List (Continued)

Additional Equipment		
MWD first aid kit	Explosives and narcotics training aides	Axes, shovels, rakes, brooms
Arm protector, sleeve (hidden)	Parasite preventative mediation	Cooling vest (MWD)
Soldier first aid kit	Light set	Paw pad protectors (2 sets)
Bite suit (optional)	Generator	Foot lockers (2) or blood boxes
Dog shampoo	Transformer	Equipment duffel bag
Heaters and fans	Warning signs (MWD)	Camouflage netting with poles

Appendix F

Force Protection Conditions

The FPCONs outlined below describe the progressive level of a terrorist threat to all US military facilities and personnel under *DOD Directive 2000.12*. As approved by the Joint Chiefs of Staff, the terminology and definitions in this appendix are recommended security measures designed to ease interservice coordination and to support US military antiterrorism activities. Each installation must localize these measures for its specific circumstances, accounting for service and commander requirements, local laws, and SOFAs as a minimum. See *AR 525-13* for further details concerning FPCONs. This appendix details what each condition is and the suggested actions of the MWD teams for each level and measure. Commanders are encouraged to employ MWD teams wherever an MWD team can enhance security and force protection measures that are already established.

FORCE PROTECTION CONDITION NORMAL

F-1. FPCON normal applies when a general global threat of possible terrorist activity exists. It warrants a routine security posture.

FORCE PROTECTION CONDITION ALPHA

F-2. FPCON Alpha applies when there is an increased general threat of possible terrorist activity against personnel or facilities, the nature and extent of which are unpredictable. Alpha measures must be capable of being maintained indefinitely. The following bullets explain the FPCON Alpha measures:

- **Alpha measure 1.** Inform personnel and family members of the general situation at regular intervals. Also, remind them to be alert for and report suspicious activities such as the presence of unfamiliar personnel and vehicles, suspicious/unattended parcels, and possible surveillance attempts.
- **Alpha measure 2.** Ensure that the duty officer or personnel with access to building plans as well as the plans for area evacuations are available at all times. Plans to execute access control procedures should be in place. Key personnel required to implement security plans should be on call and readily available.
- **Alpha measure 3.** Secure and randomly inspect buildings, rooms, and storage areas not in regular use.

Note: MWD should randomly inspect buildings after normal duty hours. The KM should ensure that no inspection time or location patterns are ever set by MWD teams.

- **Alpha measure 4.** Increase security spot checks of vehicles and persons entering installations under the jurisdiction of the United States.

Note: MWD should perform vehicle checks randomly at entry points.

- **Alpha measure 5.** Limit access points for vehicles and personnel commensurate with a reasonable flow of traffic.

- **Alpha measure 6.** Apply Alpha measures 4, 5, 7, or 8 either individually or in combination with each other as a deterrent.
- **Alpha measure 7.** Review intelligence, counterintelligence, and operations dissemination procedures.
- **Alpha measure 8.** Review and, if necessary, implement security measures for HRP.
- **Alpha measure 9.** Consult local authorities on the threat and mutual antiterrorism measures as appropriate.
- **Alpha measure 10.** Review all plans and be prepared to implement a higher FPCON.
- **Alpha measure 11 (spare).** Add additional measures as deemed appropriate by the commanders.

FORCE PROTECTION CONDITION BRAVO

F-3. FPCON Bravo applies when an increased or more predictable threat of terrorist activity exists. Sustaining Bravo measures for a prolonged period may affect operational capabilities and impact relations with local authorities. The following bullets explain the FPCON Bravo measures:

- **Bravo measure 1.** Continue, or introduce, all measures in the previous FPCON level.
- **Bravo measure 2.** Check plans for the implementation of the next FPCON level.
- **Bravo measure 3.** Identify critical and high-occupancy buildings. Move cars and objects (such as crates and trash containers) away from buildings to reduce vulnerability to bomb attacks. Consider centralized parking.
- **Bravo measure 4.** Secure and inspect all buildings, rooms, and storage areas not in regular use.
- **Bravo measure 5.** Inspect the interior and exterior of buildings in regular use for suspicious packages at the beginning and end of each workday, as well as at random intervals. Patrol teams search the exterior of high-occupancy buildings at the beginning and end of each day.

Note: Use PEDD to search buildings and areas.

- **Bravo measure 6.** Implement mail-screening procedures to identify suspicious letters and parcels.

Note: Search the mail distribution center with both PEDD and PNDD teams randomly.

- **Bravo measure 7.** Inspect commercial deliveries randomly. Advise family members to check home deliveries.
- **Bravo measure 8.** Increase presence around and surveillance of domestic accommodations, schools, messes, clubs, and other soft targets to improve deterrence and defense and build confidence among staff and dependents.

Note: Randomly patrol and perform walking patrols of populated areas with MWDs.

- **Bravo measure 9.** Implement plans to enhance off-installation security of DOD facilities. In areas with threat levels of moderate, significant, or high, coverage includes facilities (such as DOD schools and daycare centers) and transportation services and routes (such as bus routes) used by DOD employees and family members.
- **Bravo measure 10.** Inform local security committees of actions being taken.
- **Bravo measure 11.** Verify the identity of visitors and randomly inspect their suitcases, parcels, and other containers.
- **Bravo measure 12.** Conduct random patrols to check vehicles, people, and buildings.
- **Bravo measure 13.** Implement additional security measures for HRP.
- **Bravo measure 14.** Identify and brief personnel who may augment guard forces. Review specific rules of engagement including the use of deadly force.

- **Bravo measure 15.** Test the mass notification system.
- **Bravo measure 16.** Verify the identity of all personnel entering buildings as deemed appropriate.
- **Bravo measure 17.** Identify off-installation safe havens for use at the next FPCON.
- **Bravo measure 18.** Review and implement proper OPSEC, communications security (COMSEC), and information security (INFOSEC) status.
- **Bravo measure 19.** Consider using tactical deception techniques.
- **Bravo measure 20 (spare).** Add additional measures as deemed appropriate by the commanders.

FORCE PROTECTION CONDITION CHARLIE

F-4. FPCON Charlie applies when an incident occurs or intelligence is received indicating that some form of terrorist action or targeting against personnel or facilities is likely. Implementing FPCON Charlie measures creates hardship and affects the activities of the unit and its personnel. The following bullets explain the FPCON Charlie measures:

- **Charlie measure 1.** Continue, or introduce, all measures in the previous FPCON.
- **Charlie measure 2.** Ensure that duty personnel are cognizant of actions required to implement antiterrorism and force protection plans. Recall required personnel.
- **Charlie measure 3.** Reduce access control to a minimum and strictly enforce entry. Randomly search vehicles.

Note: Increase the PEDD and PNDD teams' search times at entry points.

- **Charlie measure 4.** Increase standoff from sensitive buildings based on the threat.
- **Charlie measure 5.** Issue weapons and appropriate communication equipment to guards according to SOFAs. Include specific orders on the issue of ammunition.
- **Charlie measure 6.** Increase patrols of the installation to include waterside perimeters, if appropriate.

Note: Use MWD assets to expand the perimeter patrol area and detect intruders.

- **Charlie measure 7.** Implement a barrier plan to prevent a vehicle-borne attack.
- **Charlie measure 8.** Protect all designated vulnerable points. Give special attention to vulnerable points outside the military establishment.
- **Charlie measure 9.** Consult local authorities about closing public (and military) roads and facilities to reduce the vulnerability to attack.
- **Charlie measure 10.** Consider searching suitcases, briefcases, and packages being brought onto the installation through ACPs and consider randomly searching suitcases, briefcases, and packages leaving.

Note: Use PEDD and PNDD teams to search the items above.

- **Charlie measure 11.** Review personnel policy procedures to determine a course of action for family members.
- **Charlie measure 12.** Review access procedures for all non-US personnel and adjust as appropriate.
- **Charlie measure 13.** Consider escorting children to and from DOD schools.
- **Charlie measure 14 (spare).** Add additional measures as deemed appropriate by the commanders.

FORCE PROTECTION CONDITION DELTA

F-5. FPCON Delta applies in the immediate area where a terrorist attack has occurred or when intelligence has been received that terrorist action against a specific location or person is imminent. Normally, FPCON Delta is declared as a localized condition. The following bullets explain the FPCON Delta measures:

- **Delta measure 1.** Continue, or introduce, all measures in the previous FPCON.
- **Delta measure 2**. Augment guards as necessary.
- **Delta measure 3.** Identify all vehicles within operational or mission support areas.
- **Delta measure 4.** Search all vehicles and contents before allowing entrance to the installation.

Note: Use PEDD and PNDD as needed to search vehicles and contents.

- **Delta measure 5.** Control access and implement positive identification of all personnel with no exceptions.
- **Delta measure 6.** Search all suitcases, briefcases, and packages brought into the installation.

Note: Use PEDD and PNDD as needed to search the items above.

- **Delta measure 7.** Escort children to and from DOD schools or close the schools.
- **Delta measure 8.** Make frequent checks of the exterior of buildings and of parking areas. Use MWD teams to make frequent checks of the exterior of buildings and parking areas.
- **Delta measure 9.** Restrict all nonessential movement.
- **Delta measure 10.** Close public and military roads and facilities if permitted.
- **Delta measure (spare).** Add additional measures as deemed appropriate by the commanders.

Appendix G

After-Action Reviews

The AAR process is used to document and discuss MWD operations and training. AARs are focused on performance standards that enable other MWD handlers, training NCOs, KMs, program managers, and concerned agencies to discover what happened, why it happened, and how to sustain strengths and improve on weaknesses. Leaders, planners, and developers can use AARs to get the maximum benefit from every mission or task conducted during predeployment, deployment, and redeployment.

AFTER-ACTION REVIEW PURPOSE

G-1. AARs must provide—

- Candid insights into specific soldier, leader, and MWD strengths and weaknesses from various perspectives.
- Feedback and insight critical to future successes.
- Details that evaluation reports alone often lack.

G-2. The USAMPS MWD Career Management NCO collects historical data, end-of-tour reports, and other organizational specific issues. With the submission of AARs, USAMPS can build a historical database to benefit all in the improvement of Army MWD team tactics, techniques, and procedures. Ultimately, information will be submitted to the Center for Army Lessons Learned (CALL) for distribution to the Army.

AFTER-ACTION REVIEW SUBMISSION

G-3. After the completion of TDY missions, special operations, deployments or other significant events, MWD handlers submit AARs to their respective MACOMs. MACOMs forward the AARs to HQDA Office of the Provost Marshal General (OPMG) and USAMPS.

POINT OF CONTACT

G-4. The point of contact for the MWD Career Management NCO is commercial: (573) 563-8040 and defense switched network (DSN): 676-8040/7946. Send hard-copy correspondence to: US Army Military Police School, ATTN: ATSJ-P, 401 MANSCEN Loop, Suite 1058, Fort Leonard Wood, MO 65473-8926.

EXAMPLE AFTER-ACTION REVIEW

G-5. *Figure G-1*, page G-2, should be used as the format for AARs submitted to the USAMPS MWD Career Management NCO. Refer to *TC 25-20* for comprehensive guidance on AARs.

DEPARTMENT OF THE ARMY
42^{D} Military Police Detachment
Ft. Bragg, NC 27380

OFFICE SYMBOL DATE

MEMORANDUM FOR RECORD

SUBJECT: After-Action Review, Task Force Iron Dog, Sep 03 – Nov 03

1. State what was supposed to happen: This paragraph contains a brief introduction of the mission that was being performed by the MWD team or section. Include operation objectives, the overall mission, and the intent from the task force commander as well as any other relevant information.

2. State what happened: This paragraph contains pertinent information about what actually happened during the mission. Was the mission accomplished as stated above or were other objectives established inside of the area of operation?

3. Determine what went right and wrong during the mission:

   a. Topic: What was right and wrong with what happened during the mission.

   b. Discussion: Give the who, what, when, where, and how for the mission. Give a good summary of what the issue was.

   c. Recommendation: Give recommendations to improve the issues that were identified as needing improvement.

4. Point of contact for this memorandum is the undersigned at DSN: 123-4567 or jon.doe@us.army.mil.

Jon Doe
SSG, USA
Kennel Master / PEDD Handler

Note: If you have a positive topic just use Topic and Discussion sections.

Figure G-1. Sample AAR Format

Appendix H

Health and Welfare Inspections Briefing Guide

Commanders frequently request MWD support during health and welfare inspections. Commanders should be made aware that the conduct and responsibility of health and welfare inspections lies within their purview and that the MWD team is not the authority for the mission. This appendix identifies the actions that need to be addressed with the commander to ensure a successful and lawful health and welfare inspection. These actions are needed to establish and act on probable cause determinations. Questions concerning these matters should be directed to the local Staff Judge Advocate Office for clarification.

COORDINATION

H-1. The KM coordinates and approves each health and welfare inspection. After the KM receives the request for assistance, he assesses the needs of the mission and determines which detector dog team (if any) will assist the commander. The KM notifies the unit commander or 1SG and accepts or denies the mission. Once the mission is accepted, the KM briefs them on their roles and responsibilities and informs them that if his criteria are not met, the KM will terminate the search.

HANDLERS BRIEFING

H-2. Prior to the inspection, the KM, plans NCO, or handler ensures that the commander requesting support is aware that the building being inspected must be unoccupied and that all doors are unlocked (or keys available on site). On the arrival of the MWD team at the location of the health and welfare inspection, the handler performs a commander's briefing and demonstration. The handler—

- Explains the probable cause folder to the commander.
- Shows the commander the training aids used for detection.
- Demonstrates the dog's ability to detect.

H-3. A briefing document shows handlers how to record the details of each health and welfare inspection (*Figure H-1*, page H-3). The commander is provided with a copy of the briefing document containing the following information:

- The MWD teams used for the inspection.
- The canines' names and tattoo numbers.
- The scope of the proficiency training of the MWD teams. Discuss with the commander the standards and scope of the detector training required to maintain the certification of assigned MWD teams.
- The MWDs' accuracy rates. Explain to the commander that the accuracy percentage rate is based on quarterly training.
- The MWDs' detection experience. Inform the commander of the amount of time that the MWDs have actually spent conducting detection missions. Include any possible finds in prior inspections.
- The MWDs' capabilities. Explain to the commander that the MWDs are capable of detecting either drugs or explosives.

- The MWDs' alert response. Describe what the MWDs will do when they indicate the presence or odor of found drugs or explosives.
- The unit representatives. Notify the commander that one SSG or above is required to escort each MWD team through out the inspection. The commander may not be a unit representative.

UNIT REPRESENTATIVE DUTIES AND RESPONSIBILITIES

H-4. Ensure that the unit representatives are informed that they—

- Escort the MWD team at all times.
- Unlock doors or areas as needed.
- Carefully observe the MWD team.
- Follow the handler's instructions.
- Obey safety precautions.
- Spot areas that are missed by the MWD team.

DOCUMENTATION

H-5. It is important that the MWD handler documents the commander's briefing on every occasion. *Figure H-1*, page H-3, gives an example for documenting the briefing given to the commander before a health and welfare inspection. Advise the commander to maintain the handlers briefing and other documentation concerning each response of the MWD. Advise the commander that a positive response by a narcotic detector dog provides probable cause for a command-directed urinalysis. Ensure that the commander signs the briefing form and then provide a copy for filing in case of a judicial action.

INSPECTION

H-6. After the briefing has been conducted, the inspection may begin. Remember to advise the commander and unit representatives that this health and welfare inspection is the unit's mission and that the MWD team is assisting in the mission. The MWD team is not in charge of the health and welfare inspection but can be used as a tool by the commander.

ATZP-PM-K DATE: ____________

SUBJECT: Commander's Briefing for Health and Welfare Inspections Utilizing Detector Dogs

1. (commander's name) has requested narcotics detector dog support for a health and welfare inspection on (dd/mm/yyyy) .

2. Prior to canine(s) being utilized, the building must be unoccupied and all doors must be unlocked (or keys available).

3. The following provides the commander with information essential to a successful inspection:

_____ A. Team(s) to be utilized: (number of PEDD and PNDD to be used) .

_____ B. MWD name(s) and tattoo number(s): (name and tattoo number for each MWD)

_____ C. The scope of formal training of the canine team(s). Discuss detector training. Discuss the standards that must be successfully completed to maintain certification of detector dog team(s).

_____ D. The canine's accuracy rate(s) is/are (accuracy rates for each MWD) . This is based on the previous 3 months training records.

_____ E. The extent of actual MWD detection field experience (experience for each MWD)

_____ F. Explain the types of drugs or explosives that the canine(s) is/are trained to detect.

_____ G. Describe the canine's specific behavior(s)/actions when it responds, indicating the presence or the odor of narcotics (detector response of each MWD) .

_____ H. Ensure that the commander(s) is/are aware that one representatives (SSG or above) is needed to act as an escort for each team. The commander may not be one of the individuals accompanying canine team(s).

_____ I. Ensure that the escorts are briefed on their duties and responsibilities. Discuss the safety precautions that they must follow, and ensure the safety of others.

_____ J. Ensure that the commander(s) is/are aware that the unit is responsible for maintaining proper documentation (date, time, room number, common or personal area and occupant[s] name[s]), concerning the response of the canine(s). This information should be kept on file with the commander's copy of this briefing sheet, in case of a judicial action.

_____K. Ensure that the commander is aware that a positive response by a narcotic detector dog is probable cause for a command-directed urinalysis on the occupant or occupants of that room.

I (commander's name) have received a personal briefing covering all the information stated above and now understand the capabilities of the canine team(s). I am satisfied that the canine team(s) have the training and experience necessary to provide reliable information upon which to authorize a search and/or perform a command-directed urinalysis. I authorize the canine team(s) to inspect the following area(s): (areas to be inspected)

Figure H-1. Health and Welfare Briefing Document Sample

HANDLER(S) SIGNATURE

1. ______________________ 2. ______________________

3. ______________________ 4. ______________________

COMMANDER'S INFORMATION	**1SG'S INFORMATION**
NAME: ______________	NAME: ______________
RANK: ______________	RANK: ______________
UNIT: ______________	UNIT: ______________
SIGNATURE: ______________	SIGNATURE: ______________
UNIT REPRESENTATIVE	**UNIT REPRESENTATIVE**
NAME: ______________	NAME: ______________
RANK: ______________	RANK: ______________
UNIT: ______________	UNIT: ______________
SIGNATURE: ______________	SIGNATURE: ______________

Figure H-1. Health and Welfare Briefing Document Sample (Continued)

Appendix I

Terrorist Bomb Threat Stand-off Distances

Table I-1 contains information from the DOD technical support working group. Commanders, PMs, and MWD planners can use this information as a guide to determine the minimum distances necessary to evacuate personnel from suspected explosive threats.

Table I-1. Minimum Distances for Personnel Evacuation

Threat Description	*Explosives Capacity* (TNT Equivalent)*	*Building Evacuation Distance***	*Outdoor Evacuation Distance****
Pipe bomb	5 lbs/2.3 kg	70 ft/21 m	850 ft/259 m
Briefcase/suitcase bomb	50 lbs/23 kg	150 ft/46 m	1,850 ft/564 m
Compact sedan	500 lbs/227 kg	320 ft/98 m	1,500 ft/457 m
Sedan	1,000 lbs/454 kg	400 ft/122 m	1,750 ft/534 m
Passenger/cargo van	4,000 lbs/1,814 kg	640 ft/195 m	2,750 ft/838 m
Small moving van/delivery truck	10,000 lbs/4,536 kg	860 ft/263 m	3,750 ft/1,143 m
Moving van/water truck	30,000 lbs/13,608 kg	1,240 ft/375 m	6,500 ft/1,982 m
Semitrailer	60,000 lbs/27,216 kg	1,570 ft/475 m	7,000 ft/2,134 m

* Based on maximum volume or weight of explosives (TNT equivalent).

** Governed by the ability of an unstrengthened building to withstand severe damage or collapse.

*** Governed by either the fragment throw distance or glass breakage/falling glass hazard distance (whichever throws fragments farther than vehicle bombs).

NOTES:

1. All personnel must either seek shelter inside a building away from windows and exterior walls (with some risk) or move beyond the outdoor evacuation distance.
2. The preferred area is beyond the outdoor evacuation distance for the evacuation of personnel in buildings. This is mandatory for personnel outdoors.

Glossary

Acronyms and Abbreviations

1SG	first sergeant
AAR	after-action review
ACP	access control point
AF	Air Force
agg	aggression
AR	Army regulation
ASG	area support group
ASI	additional skill identifier
ASP	ammunition supply point
ATTN	attention
BDU	battle dress uniform
C2	command and control
C4	composition 4
CA	corrective action
CALL	Center for Army Lessons Learned
CCTV	Closed-circuit television
cert	certification
CI	civilian internee
CID	Criminal Investigation Division
clg	cleaning
CMFO	centralized furnishings management office
comm	communications
COMSEC	communications security
CONEX	container express
cont	continuted
CP	command post
CPT	captain
CR	corrective response
DA	Department of the Army
d	deep
DARE	Drug Abuse Resistance Education
DCSPER	Dputy Chief of Stagg for Personnel
DC	detonating cord
DD	Department of Defense
DEA	Drug Enforcement Administration

Dec	December
det	detonating
DF	deficiency
DOD	Department of Defense
DODMWDVS	Department of Defense Military Working Dog Veterinary Services
DSN	defense switched network
DST	drug suppression team
EAC	echelon above corps
EOD	explosive ordnance disposal
EPW	enemy prisoner of war
eval	evaluaiton
FM	field manual
FMO	Furnishing managemetn office
FPCON	force protection condition
ft	foot; feet
GE	Germany
GOV	government owned vehicle
GSA	General Services Administration
h	high
HF	hole, variable
HRP	high-risk personnel
HRT	high-risk target
HVAC	heating, ventilation, and air conditioning
I/R	internment and resettlement
IDS	Intrusion Detection System
IED	improvised explosive device
INFOSEC	information security
IPB	intelligence preparation of the battlefield
iso	isolation
kg	kilogram
KM	kennel master
LAFB	Lackland Air Force base
LAN	local area network
LBS	pounds
LBV	load-bearing vest
LCE	load-carrying equipment
LED	light-emitting diode
LIN	line item number
LTC	lieutenent colonel
m	meter

MACOM	major Army command
MANSCEN	Maneuver Support Center
MCI	military customs inspector
MCIPM	military customs inspector program manager
MCM	Manual for Courts-Martial
mech	mechanical
METL	mission-essential task list
METT-TC	mission, enemy, terrain, troops, time available, and civilian considerations
MEVA	mission-essential vulnerable areas
MILSTRIP	military standard transportation and issue procedures
min	minimum
MMSO	maneuver mobility support operations
MOPP	mission-oriented protective posture
MPR	military police report
MSR	main supply route
MTOE	modification table of organization and equipment
MWD	military working dog
MWDMS	Military Working Dog Management System
NBC	nuclear, biological, and chemical
NCO	noncommissioned officer
NCOIC	noncommissioned officer in charge
no	number
Nov	November
NSN	national stock number
OB	obstacle
Oct	October
ODCSOPS	Office of the Deputy Chief of Staff for Operations and Plans
off/L	off leash
on/L	on leash
OIF	Operation Iraqi Freedom
OPMG	Office of the Provost Marshal General
OPSEC	operational security
pam	pamphlet
PC	potassium chlorate
PD	patrol dog
PEDD	patrol explosive detector dog
PIO	police intelligence operations
PM	provost marshal
PMO	provost marshal office

PNDD	patrol narcotic detector dog
POV	privately owned vehicle
PT	physical training
prep	preparation
RAMP	random antiterrorism measures program
rec	recreation
S2	intelligence section
SC	sodium cholorate
sec	seconds
Sep	September
SF	standard form
SFC	sergeant first class
SGT	sergeant
SOFA	Status of Forces Agreement
SOP	standard operating procedures
SP	smokeless powder
SRT	special-reaction team
SSG	staff sergeant
STX	Semtex
TDA	table of distribution and allowancees
TDY	temporary duty
TISA	troop issue subsistence activity
TMP	transportation motor pool
TNT	trinitrotoluene
TOE	table of organization and equipment
TRADOC	United States Army Training and Doctrine Command
Tues	Tuesday
US	United States
USA	United States of America
USC	United States Code
USCS	United States Customs Service
USSS	United States Secret Service
v	version
VCO	veterinary corps officer
vet	veterinarian
VIP	very important person
VTF	Veterinary Treatment Facility
VTR	Veterinary Treatment Record
WT	weight

References

DA Forms are available on the Army Publishing Directorate web site <http://www.apd.army.mil>. DD forms are available from the Office of the Secretary of Defense web site <http://www.dtic.mil/whs/directives/infomgt/forms/formsprogram.htm>.

SOURCES USED

These are the sources quoted or paraphrased in this publication.

AR 190-12. *Military Working Dogs.* 30 September 1993.

AR 190-14. *Carrying of Firearms and Use of Force for Law Enforcement and Security Duties.* 12 March 1993.

DA Pam 190-12. *Military Working Dog Program.* 30 September 1993.

DA Pam 611-21. *Military Occupational Classification and Structure.* 31 March 1999.

DOCUMENTS NEEDED

These documents must be available to the intended users of this publication.

AR 700-81. *DOD Dog Program.* 5 May 1971.

DA Form 2807-R. *Military Working Dog Training and Utilization Record.*

DA Form 3992–R. *Narcotics or Explosives Detector Dog Training and Utilization Record.*

FM 3-19.15. *Civil Disturbance Operations.* 18 April 2005.

READINGS RECOMMENDED

These sources contain relevant supplemental information.

AF Form 1256. *Certificate of Training.*

AR 190-47. *The Army Corrections System.* 5 April 2004.

AR 190-8. *Enemy Prisoners of War, Retained Personnel, Civilian Internees and Other Detainees.* 1 October 1997.

AR 350-1. *Army Training and Education.* 9 April 2003.

AR 40-1. *Composition, Mission, and Functions of the Army Medical Department.* 1 July 1983.

AR 40-3. *Medical, Dental, and Veterinary Care.* 12 November 2002.

AR 40-905. *Veterinary Health Services.* 16 August 1994.

AR 525-13. *Antiterrorism.* 4 January 2002.

AR 600-8-101. *Personnel Processing (In-, Out-, Soldier Readiness, Mobilization, and Deployment Processing).* 15 July 2003.

AR 600-9. *The Army Weight Control Program.* 10 June 1987.

AR 614-200. *Enlisted Assignments and Utilization Management.* 30 September 2004.

DA Form 2028. *Recommended Changes to Publications and Blank Forms.*

DA Form 3161. *Request for Issue or Turn-In.*

DA Form 5513-R. *Key Control Register and Inventory.*

DA Form 6. *Duty Roster*

DA Form 7281-R. *Command Oriented Arms, Ammunition, & Explosives Security Screening and Evaluation Record.*

DA Pam 710-2-1. *Using Unit Supply System (Manual Procedures).* 31 December 1997.

DD Form 1834. *Military Working Dog Service Record.*

DEA Form 223. *Controlled Substance Registration Certificate.*

DOD Directive 2000.12. *DoD Antiterrorism (AT) Program.* 18 August 2003.

DOD Directive 5030.49. *DoD Customs and Border Clearance Program.* 4 May 2004.

DOD Directive 6400.4. *DoD Veterinary Services Program.* 22 August 2003.

DOD Instruction 3025.12. *Military Assistance for Civil Disturbances (MACDIS).* 4 February 1994

FM 21-20. *Physical Fitness Training.* 30 September 1992.

FM 27-10. *The Law of Land Warfare.* 18 July 1956.

FM 3-19.40. *Military Policy Internment/Resettlement Operations.* 1 August 2001.

Lackland AFB Form 375. *Patrol Dog Certification.*

Lackland AFB Form 375a. *Detector Dog Certification.*

MCM, Part III Military Rules of Evidence; Rule 311, *Evidence obtained from unlawful searches and seizures.* 2000.

MCM, Part III Military Rules of Evidence; Rule 313, *Inspections and inventories in the armed forces.* 2000.

MCM, Part III Military Rules of Evidence; Rule 314, *Searches not requiring probable cause.* 2000.

MCM, Part III Military Rules of Evidence; Rule 315, *Probable cause searches.* 2000.

MCM, Part III Military Rules of Evidence; Rule 315(f)(2), *Probable cause determination.* 2000.

MCM, Part III Military Rules of Evidence; Rule 316, *Seizures.* 2000.

SF 702. *Security Container Check Sheet.*

TC 19-210. *Access Control Handbook.* 4 October 2004.

TC 25-20. *A Leader's Guide to After Action Reviews.* 30 September 1993.

USC Title 18, Crimes and Criminal Procedure; Part I, Crimes; Section 1385, *Use of Army and Air Force as posse comitatus.*

Index

Made in the USA
Middletown, DE
05 November 2026

36587662R00136